SITE AUTOMATION

Automated/Robotic On-site Factories

The Cambridge Handbooks on Construction Robotics series focuses on the implementation of automation and robot technology to renew the construction industry and to arrest its declining productivity. The series is intended to give professionals, researchers, lecturers, and students basic conceptual and technical skills and implementation strategies to manage, research, or teach the implementation of advanced automation and robot-technology-based processes and technologies in construction. Currently, the implementation of modern developments in product structures (modularity and design for manufacturing), organizational strategies (just in time, just in sequence, and pulling production) and informational aspects (computer-aided design/manufacturing or computer-integrated manufacturing) are lagging because of the lack of modern integrated machine technology in construction. The Cambridge Handbooks on Construction Robotics books discuss progress in robot systems theory and demonstrate their integration using real systematic applications and projections for off-site as well as on-site building production.

Site Automation – Automated/Robotic On-site Factories extends the new technology of robotics in building-component manufacturing and construction introduced in earlier volumes to on-site structured environments and on-site automated factories – for both building construction and disassembly. This volume explores 30 different worldwide systems within a careful analytical framework in which the best conceptual features are extracted to help professionals and researchers develop new applications. The analytical approach splits the systems studies into a technical portion and a portion that focuses on aspects related to productivity, efficiency, and economic performance. A benefit of automated on-site factories is the integration of several stand-alone construction robots into structured on-site environments to achieve significant improvements in on-site organization, integration, and material flow. The book outlines each system analysed in both textual and graphical form, introduces an expandable categorization system, and provides all relevant background information.

Thomas Bock is a professor in building realization and robotics at Technische Universität München (TUM). His research focuses on automation and robotics in building construction, from planning building production and utilization to reorganization and deconstruction. After his academic studies in architecture at the University of Stuttgart and the Illinois Institute of Technology, he completed his doctorate at the University of Tokyo. He is on the board of directors of the International Association for Automation and Robotics in Construction (IAARC), the Asian Habitat Society in Peking, and the International Institute of Construction Information in Tokyo. He operates as a consultant at Ministère de l'Emploi de la Cohésion Sociale et du Logement in France and is a member of the Academy of Architecture and Civil Engineering, the Petrovich Academy of Sciences, and the Academy of Informatics. Professor Bock is on the editorial boards of top journals such as *Robotica*, *Automation in Construction*, and *the International Journal of Construction Management*. Recently he became coordinator of Customized Industrial Construction's newly founded W119.

Thomas Linner is a research associate in building realization and robotics at Technische Universität München (TUM). During the last few years, he has supervised some major research projects, with a focus on the deployment of advanced technology in the building sector. He is a specialist in the area of automated production of building products as well as in the enhancement of the performance of building products by advanced technology. He completed his dissertation in the field of construction automation, focussing on automated/robotic on-site factories. Increasingly, the generation of innovation strategies, business models, value systems, and innovative manufacturing organization methods complementary with advanced technology in construction is becoming the central issue in his research. Dr. Linner has been an invited speaker at universities such as the University of Tokyo and Cambridge University. He has received several prizes and grants, including a Japanese Center of Excellence Grant for research in Japan.

CAMBRIDGE HANDBOOKS ON CONSTRUCTION ROBOTICS

Robot-Oriented Design: Design and Management Tools for the Deployment of Automation and Robotics in Construction ISBN 9781107076389

Robotic Industrialization: Automation and Robotic Technologies for Customized Component, Module, and Building Prefabrication ISBN 9781107076396

Site Automation: Automated/Robotic On-Site Factories ISBN 9781107075979

Construction Robotics: Elementary Technologies and Single Task Construction Robotics ISBN 9781107075993

Ambient Robotics: Automation and Robotic Technologies for Maintenance, Assistance, and Service ISBN 9781107075986

Site Automation

AUTOMATED/ROBOTIC ON-SITE FACTORIES

Thomas Bock
Technische Universität München

Thomas Linner
Technische Universität München

CAMBRIDGE
UNIVERSITY PRESS

CAMBRIDGE
UNIVERSITY PRESS

Shaftesbury Road, Cambridge CB2 8EA, United Kingdom

One Liberty Plaza, 20th Floor, New York, NY 10006, USA

477 Williamstown Road, Port Melbourne, VIC 3207, Australia

314–321, 3rd Floor, Plot 3, Splendor Forum, Jasola District Centre, New Delhi – 110025, India

103 Penang Road, #05–06/07, Visioncrest Commercial, Singapore 238467

Cambridge University Press is part of Cambridge University Press & Assessment, a department of the University of Cambridge.

We share the University's mission to contribute to society through the pursuit of education, learning and research at the highest international levels of excellence.

www.cambridge.org
Information on this title: www.cambridge.org/9781107075979

First published 2016

A catalogue record for this publication is available from the British Library

Library of Congress Cataloging-in-Publication data
Bock, Thomas, 1957–
Logistics, site automation, and robotics : automated and robotic on-site factories / Thomas Bock, Technische Universität München, Thomas Linner, Technische Universität München.
pages cm
Includes bibliographical references and index.
ISBN 978-1-107-07597-9 (hardback)
1. Industrialized building. 2. Construction industry – Automation.
3. Robotics. I. Linner, Thomas, 1979– II. Title.
TH1000.B63 2014
338.4′5624–dc23 2014020956

ISBN 978-1-107-07597-9 Hardback

Contents

Acknowledgements

Construction automation gained momentum in the 1970s and 1980s in Japan, where the foundations for real-world application of automation in off-site building manufacturing, single-task construction robots, and automated construction sites were laid. This book series carries on a research direction and technological development established within this "environment" in the 1980s under the name *Robot-Oriented Design,* which was a focal point of the doctoral thesis of Thomas Bock at the University of Tokyo in 1989. In the context of this doctoral thesis many personal and professional relationships with inventors, researchers, and developers in the scientific and professional fields related to the construction automation field were built up. The doctoral thesis that was written by Thomas Linner (*Automated and Robotic Construction: Integrated Automated Construction Sites*) in 2013 took those approaches further and expanded the documentation of concepts and projects. Both of theses form the backbone of the knowledge presented in this book series.

The authors first wish to express their deepest gratitude to Prof. Dr. Yositika Uchida, Prof. Dr. Y. Hasegawa, and Prof. Dr. Umetani (TIT); Dr. Tetsuji Yoshida, Dr. Junichi Maeda, Dr. Yamazaki, Dr. Matsumoto, Mr. Abe, and Dr. Ueno (Shimizu); Prof. Yashiro, Prof. Matsumura, Prof. Sakamura, Prof. Arai, Prof. Funakubo, Prof. Hatamura, Prof. Inoue, Prof. Tachi, Prof. Sato, Prof. Mitsuishi, Prof. Nakao, and Prof. Yoshikawa (UoT); Dr. Oiishi and Mr. Fujimura (MHI); Dr. Muro, Kanaiwa, Miyazaki, Tazawa, Yuasa, Dr. Sekiya, Dr. Hoshino, Mr. Arai, and Mr. Morita (Takenaka); Dr. Arai, Dr. Chae, Mr. Mashimo, and Mr. Mizutani (Kajima); Dr. Shiokawa, Dr. Hamada, Mr. Furuya, Mr. Ikeda, Mr. Wakizaka, Mr. Suzuki, Mr. Kondo, Mr. Odakuda, Mr. Harada, Mr. Imai, Mr. Miki, and Mr. Doyama (Ohbayashi); Dr. Yoshitake, Mr. Sakou, Mr. Takasaki, Mr. Ishiguro, Mr. Morishima, Mr. Kato, and Mr. Arai (Fujita); Mr. Nakamura, Mr. Tanaka, and Mr. Sanno (Goyo/Penta Ocean); Mr. Nagao and Mr. Sone (Maeda); Mr. Ohmori, Mr. Maruyama, and Mr. Fukuzawa (Toda); Dr. Ger Maas (Royal BAM Group); Dr. Espling, Mr. Jonsson, Mr. Fritzon, Mr. Junkers, Mr. Andersson, and Mr. Karlson (Skanska); Mrs. Jansson, Mrs. Johanson, Mr. Salminen, Mr. Apleberger, Mr. Lindström, Mr. Engström, and Mr. Andreasson (NCC); Mr. Hirano (Nishimatsu); Mr. Weckenmann (Weckenmann); Mr. Ott and Mr. Weinmann (Handtech); Mr. Bauer and Mr. Flohr (Cadolto); Mr. Kakiwada, Mr. Ohmae, Mr. Sawa, Mr. Yoshimi, Mr. Shimizu, and Mr. Oki (Hazama Gumi); Mr. Masu, Mr. Oda, and Mr. Shuji (Lixil); Mr. Yanagihara, Mr. Oda, Mr. Kanazashi, and

Mr. Mitsunaga (Tokyu Kensetsu); Mr. Okada (Panasonic); Mr. Nakashima and Mr. Kobayashi (Kawasaki HI); Mr. Tazawa (Ishikawajima HI); Mr. Okaya (IHI Space); Dr. Morikawa and Mr. Shirasaka (Mitsubishi Denki); Mr. Maeda, Mr. Matsumoto, and Mr. Kimura (Hitachi Zosen); Mr. Shiroki and Mr. Yoshimura (Daiwa House); Mr. Hagihara and Mr. Hashimoto (Misawa Homes); Mr. Okubo (Mitsui Homes); Dr. Itoh, Mr. Shibata, Mr. Kasugai, Mr. Kato, and Mr. Komeyama (Toyota Homes); Mr. Hirano, Mr. Kawano, and Mr. Takahashi (Toyota Motors); Dr. K. Ohno (+); Dr. Misawa, Dr. Mori, Nozoe, Tamaki, Okada, and Dr. Fujii (Taisei); Prof. P. Sulzer and Prof. Helmut Weber (+); Prof. Dr. Fazlur Khan (+); Prof. Dr. Myron Goldsmith (+ IIT); Prof. A. Warzawski (+ Technion); Prof. R. Navon, Prof. Y. Rosenfeld, Prof. Dr. S. Isaac, Prof. A. Mita, Prof. A. Watanabe, Prof. Kodama, and Prof. Suematsu (Toyota National College of Technology); Mr. Suketomo, Mr. Yoshida, Mr. Oku, and Mr. Okubo (Komatsu); Dr. Ogawa and Mr. Tanaka (Yasukawa); Dr. Inaba (Fanuc); Mr. Usami (Hitachi RI); Mr. Ito and Mr. Hattori (Mitsubishi RI); Mr. Noda, Mr. Ogawa, and Mr. Yoshimi (Toshiba); Mr. Senba (Hitachi housetec); Mr. Sakurai, Mr. Hada, Mr. Sugiura, Mr. Yoshikawa, Mr. Shimizu, Mr. Tanaka, Mr. Nishiwaki, and Mr. Nomura (Sekisui Heim); Mr. Aoki, Mr. Watanabe, Mr. Kotani, and Mr. Kudo (Sekisui House); Mr. Takamoto, Mr. Mori, Mr. Baba, Mr. Sudoh, Mr. Fujita, and Mr. Hiyama (TMsuk); Prof. Dr. T. Fukuda, Prof. Iguchi, Prof. Ohbayashi, Prof. Tamaki, Mr. Yanai, and Mr. Hata (JRA); Mr. Uchida (JCMA); Dr. Kodama (Construction Ministry Japan); Dr. Sarata (AIST); Mr. Yamamoto (DoKen); Mr. Yamanouchi (Ken-Ken); Mr. Yoshida (ACTEC); Prof. Dr. Wolfgang Bley, Prof. Dr. G. Kühn, and Jean Prouve (+); Dr. Sekiya, Dr. Yusuke Yamazaki, Prof. Dr. Ando, Prof. Dr. Seike, Prof. Dr. J. Skibniewski, Prof. Dr. C. Haas, Dr. R. Wing, Prof. Dr. David Gann, Prof. Dr. Puente, Prof. Dr. P. Coiffet, Prof. Dr. A. Bulgakow, Prof. Dr. Arai, Prof. Kai, Prof. Dr. A. Chauhan, Prof. Dr. Spath, Dr. M. Hagele, Dr. J.L. Salagnac, and Prof. Cai (HIT); Prof. Dr. Szymanski (+); Prof. Dr. F. Haller (+); Prof. Dr. Tamura and Prof. Dr. Han.

The authors wish to express their very sincere gratitude to the Japanese guest Professors Dr. K. Endo and Dr. H. Shimizu for supporting and advising the authors during their stays at the authors' chair in Munich. Furthermore, the authors are very thankful to all the companies in the field of component manufacturing, prefabrication, and on-site automation outlined throughout the book series for the information, analyses, pictures, and details of their cutting-edge systems and projects that they shared and provided.

The authors are indebted to the AUSMIP Consortium and the professors and institutions involved for supporting their research. In particular, the authors extend their thanks to Prof. Dr. S. Matsumura, Prof. Dr. T. Yashiro, Prof. Dr. S. Murakami, Dr. S. Chae, Prof. Dr. S. Kikuchi, Assistant Prof. Dr. K. Shibata, Prof. Dr. S.-W. Kwon, Prof. Dr. M.-Y. Cho, Prof. Dr. D.-H. Hong, Prof. Dr. H.-H. Cho, Prof. A. Deguchi, and Prof. B. Peeters and their students and assistants for supporting the authors with information and organizing and enabling a multitude of important site visits. Furthermore, the authors gratefully acknowledge the International Association for Automation and Robotics in Construction, whose yearly conferences, activities, publications, and network provided a fruitful ground for their research and a motivation to follow their research direction.

The authors are indebted in particular to W. Pan, B. Georgescu, and K. Iturralde (in particular for their support in context with the graphical representations and images in this volume), and M. Helal for their great support in the phase of completing the volume series. The authors thank T. Isobe, S. Kitamura, and Prof. T. Seike for their support and Dr.-Ing. C. Georgoulas, A. Bittner, and J. Güttler for their advice and support. In addition, the authors thank A. Liu Cheng, B. Toornstra, M. Rahaman, I. Arshad, P. Anderson, N. Agrafiotis, K. Pilatsiou, G. Anschau Rick, A. Grisales, J. Maharjan, J. F. Jacob, and F. Gonzalez.

The authors are sincerely grateful also to their numerous motivated students within the master course Advanced Construction and Building Technology who have, as part of lab or coursework, contributed to the building of the knowledge presented and who have motivated the authors to complete this book series.

Glossary

Alignment and accuracy measurement system (AAMS): An AAMS creates a feedback loop between a system that measures how accurately components are positioned and an alignment system (e.g., a motorized unit attached temporarily to the joint of a column component) that automatically moves or aligns the *component* into the desired final position.

Assembly: In this book, assembly refers to the production of higher-level components or final products out of *parts* and lower-level *components*. The process of assembly of an individual *part* or *component* to a larger system involves positioning, alignment, and fixation operations. *Upstream* processes dealing with the generation of elements for assembly are referred to as *production*.

Automated guided vehicle (AGV): Computer-controlled automatic or robotic mobile transport or logistics vehicle.

Automated/robotic on-site factory: *Structured environment* (factory or factory-like) setup at the place of *construction*, allowing *production* and *assembly* operations to be executed in a highly systemized manner by, or through, the use of machines, automation, and robot technology.

Batch size: The amount of identical or similar products produced without interruption before the manufacturing system is substantially changed to produce another product. Generally speaking, low batch sizes are related to high fixed costs and high batch sizes are related to low fixed costs.

Building component manufacturing (BCM): BCM refers to the transformation of *parts* and low-level *components* into higher-level *components* by highly mechanized, automated, or robot-supported industrial settings.

Building integrated manufacturing technology: Automation technology, microsystem technology, sensor systems, or robot technology can be directly integrated into buildings, units, or components as a permanent system. Technology used to manufacture the building can thus become a part of the building technology.

Bundesministerium für Bildung und Forschung (BMBF): Federal Ministry of Education and Research in Germany. The BMBF funds education, research, and technical development in a multitude of industrial fields.

Capital intensity: The capital intensity (also referred to as workplace cost) is calculated by dividing the capital stock (assets, devices, and equipment used to transform/manufacture the outcome) by the number of employees in the industry.

Chain-like organization: In a chain-like organization, the *flow of material* between individual workstations is highly organized and fixed, and a material transport system linking the stations exists.

Climbing system (CS): Automated/robotic on-site factories require, especially in the manufacturing of vertically oriented buildings, a system that allows the *sky factory* (SF) to rise to the next floor, once a floor level has been completed. Most SFs, therefore, rest on stilts that transfer the loads of the SF to the building's bearing structure, or to the ground. Other CSs are able to climb along a central core, pushing up the building or, in the case of the manufacture of horizontally orientated buildings, for example, to enable the factory to move horizontally. In some cases, in addition to climbing, CSs are used to provide a fixture or template for the positioning of components by *manipulators*. Because of the enormous forces necessary to lift SFs, hydraulic systems and screw presses are common actuation systems.

Closed loop resource circulation: Systems for avoiding waste and reduction of resource consumption, by integrating concepts such as reverse logistics, remanufacturing, and recycling. Material or product that flows on a factory, utilization, and deconstruction level can be related back to the manufacturing system to close the loop.

Closed sky factory (CSF): *Sky factory* that completely covers and protects the workspace in *automated/robotic on-site factories*, thus allowing the installation of a fully *structured environment* that erases the influence of parameters that cannot be 100% specified (e.g., rain, wind).

Component: In this book, in a hierarchical modular structure, components can be divided into lower-level components and higher-level components. Components consist of subelements of *parts* and lower-level components. Higher-level components can be assembled into *modules* and *units*.

Component carriers: Component carriers and pallets (special types of component carriers) play an important role in logistics. In many cases, parts, components, or final products cannot be directly handled or manipulated by the logistics system. Component carriers and pallets act as mediators between the handled material and the actual logistics system.

Computer-aided design/computer-aided manufacturing (CAD/CAM): From the 1980s on, the novel and highly interdisciplinary research and application field CAD/CAM was formed, which aimed at integrating computerized tools and systems from the planning and engineering field with manufacturing and machine control systems to allow for a more-or-less direct use of the digital design data for automated and flexible manufacturing. The field evolved further toward *computer-integrated manufacturing*.

Computer-aided quality management (CAQM): Control of quality by software, made possible through the linking of manufacturing systems with computer systems.

Computer-integrated manufacturing (CIM): From the 1990s on, the combined *computer-aided design/computer-aided manufacturing* approach evolved into the CIM approach. The focus was then broader and the idea was that more and more fields and tools, and also business economic issues (e.g., computer-aided forecasting or demand planning), could be integrated by computerized systems to form continuous process and information chains in manufacturing that span all value-adding nodes in the value system.

Connector system: The development of connector systems that connect complex *components* in a robust way to each other is a key element in complex products such as cars, aircraft, and buildings in particular. To support efficient *assembly*, connector systems can, for example, be compliant or plug-and-play–like. Connector systems can also be designed to support efficient disassembly, remanufacturing, or recycling.

Construction: Activities necessary to build a building on-site. Construction, in this book, is interpreted as being a manufacturing process, and accordingly buildings are seen as "products."

Cycle Time: Important on the workstation level: The cycle time refers to the time allowed for all value-adding activities performed by humans and machines at a workstation within a network of workstations.

Degree of freedom (DOF): In a serial kinematic system each joint gives the systems, in terms of motion, a DOF. At the same time, the type of joint restricts the motion to a *rotation* around a defined axis or a *translation* along a defined axis.

Depth of added value: The depth of added value (e.g., measured as a percentage of the total cost of the product) refers to the total amount of value-adding activities, and thus in general to the amount of value-adding steps, realized by the *original equipment manufacturer* (OEM) or final integrator. A high depth of added value means that a large number of value-adding activities are being realized by the OEM (e.g., Henry Ford). A low depth of added value means that a low number of value-adding activities is being realized by the OEM (e.g., Dell, Smart).

Design for X (DfX): DfX strategies aim at influencing design-relevant parameters to support production, assembly maintenance, disassembly, recycling, and many other aspects related to a product's life cycle. In this book, DfX strategies are classified into four categories: DfX related to *production/assembly*, to product function, to product end-of-life issues, and to business models. In this book, *robot-oriented design* is seen as an augmentation or extension of conventional DfX strategies, consequently aiming more at the efficient use of automation and robotic technology in all four categories.

Deutscher Kraftfahrzeug-Überwachungs-Verein (DEKRA): Major German consultant and surveyor association that evaluates technical artefacts, such as cars and buildings, and defines quality and the causes of defects.

Efficiency: Efficiency can be defined as the relationship between an achieved result and the combination of factors of production. Whereas *productivity* expresses an input-to-output ratio, with a focus on a single factor of production, efficiency considers multiple factors and their combination and interrelation. Productivity can be an indication of efficiency, and efficiency itself for economic feasibility.

End-effector: The element of machines, automation systems, or robot technology that makes contact with the object to be manipulated in *manufacturing* is called the end-effector. In most cases end-effectors are modularly separable from the base system. End-effectors have a certain degree of *inbuilt flexibility*.

Factory external logistics (FEL): FEL refers to logistics systems that connect the supply network to the factory integrating and assembling the supplied *parts*, *components*, *modules*, or *units*. FEL influences the organization of the manufacturing system, *factory internal logistics*, and the factory layout.

Factory internal logistics (FIL): FIL refers to logistics systems that manipulate *parts*, *components*, *modules*, *units*, or the finished product within a manufacturing setup or factory, for example, for the transportation between various stations. Other examples include mobile and non-rail–guided transport systems, overhead crane-type material transportation systems, fixed conveyor systems allowing a component carrier or the product itself to travel in a horizontal direction in fixed lanes, and fixed conveyor systems allowing a component carrier or the product itself to travel in a vertical direction in fixed lanes. Novel cellular logistics robots combine capabilities of unrestricted mobility with horizontal and vertical transport capabilities and can travel freely and self-organize with other systems.

Factory roof structure: Structure that allows the workspace on the construction site to be covered (and therefore to be protected from outside influences such as wind, rain, or sun) and thus creates the basis for a *structured environment*. Often used as a platform for the attachment of other subsystems, such as a *climbing system*, *horizontal delivery system*, and *overhead manipulators*.

Final integrator: In this book, a final integrator refers to an entity in a value chain or value system that integrates major components into the final product. Within the OEM model, the final integrator is called *original equipment manufacturer*.

Fixed-site manufacturing: *Off-site manufacturing* or *on-site manufacturing* system that stays at a fixed place during final *assembly*.

Floor erection cycle (FEC): Time necessary to erect and finish (including technical installations and general interior finishing) a standard floor with an *automated/robotic on-site factory*.

Flow-line organization: In a flow-line organization, individual workstations do not have a fixed *flow of material*, but a general directional *flow of material* (e.g., within a factory segment or a factory).

Flow of material: Refers to material and product streams in relation to space and time that take place during the completion of a specific product in a manufacturing system and the supply network connected to it. The *efficiency* of the flow of material is determined by the arrangement of equipment, the factory layout, and the logistics processes.

Frame and infill (F&I): F&I strategies are used in a variety of industries, including the aircraft, automotive, and building industries. The idea of an F&I strategy is to use a bearing frame structure as a base element that is subsequently equipped with *parts*, *components*, systems, *modules*, and so forth during the manufacturing process. The frame thus functions as a *component carrier*. In the aircraft industry, the fuselage is interpreted as such a frame; in the automotive industry it is the car body or chassis; and in the building industry it can be seen, for example, in the form of two-dimensional (e.g., Sekisui House) or three-dimensional steel frames (Sekisui Heim).

Ground factory (GF): *Structured environment* (factory or factory-like) setup on the construction site on the ground level of the building as part of an *automated/robotic on-site factory*.

Group-like organization: In a group-like organization, individual workstations are bound together in groups. Those groups can refer to workstations with similar means of production or to workstations with complementary *means of production*. The *flow of material* between those groups can be either fixed or flexible.

Horizontal delivery system (HDS): System that transports, positions, and/or assembles *parts/components* on the construction site on a floor level.

Idle time: The unproductive standstill of a machine from end of completion to the beginning of the processing of the next material. Bottleneck operations, for example, may – when workstations are directly connected without a buffer – lead to material having to wait for a certain time until the next material can be processed and to an unproductive standstill of other workstations that are faster in processing the material.

Inbuilt flexibility: The changes in a manufacturing system can be realized without major physical or modularization enabled changes (e.g., exchange of systems, workstations, robots, *end-effectors*), but by reprogramming the existing system instead. A standard robot with 6 *degrees of freedom* (6-DOF robot) with an end-effector for welding, for example, has a high degree of flexibility and can be reprogrammed for a huge variety of welding operations within a given workspace.

Joint of a manipulator: A *manipulator* consists of at least one kinematic pair consisting of two rigid bodies (links) interconnected with a joint. The following types of joints can be distinguished: revolute joint, prismatic joint, and spherical joint.

Just in sequence (JIS): Various *parts*, *components*, and *products* are delivered from *upstream* to *downstream* workstations in the sequence in which they are handled or processed when they reach the *downstream* work stations. JIS can be performed internally within a factory or in relation to a supplier of an *original equipment manufacturer* (OEM). JIS is in most cases closely connected to *just in time* (JIT).

Just in time (JIT): Stocks and buffers are eliminated, and *parts*, *components*, and products are delivered from *upstream* to *downstream* workstations at the right time and at the right quantity. JIT can be performed internally within a factory or in relation to a supplier of an *original equipment manufacturer*. JIT is in most cases closely connected to *just in sequence*.

Kinematic base body: The combination of links and joints forms kinematic bodies that allow basic manipulation operations within a geometrically definable work space (e.g., Cartesian *manipulator*, gantry *manipulator*, cylindrical *manipulator*, spherical *manipulator*). Those kinematic base bodies consider mainly the first three axes, and thus refer mainly to *positioning* activity. For *orientation*, further *degrees of freedom* and kinematic combinations can be added on top of those base bodies.

Kinematics: Kinematics focuses on the study of geometry and motion of automated and robotic systems. It describes parameters such as position, velocity, and acceleration of joints, links, and *tool centre points* to generate mathematical models creating the basis for controlling the actuators and for finding optimized trajectories for the motions of the system. *Manipulators* are a kinematic system consisting of a multitude of kinematic subsystems, of which the kinematic pair is the most basic entity.

Large-scale prefabrication (LSP): *Off-site manufacturing* of high-level *components*, *modules*, or *units* in very large quantity by a production-line–based, automation and robotics-driven factory or factory network, interconnected in an *OEM-like integration structure*.

Link of a manipulator: A *manipulator* consists of at least one kinematic pair, consisting of two rigid bodies (links) interconnected with a *joint*.

Logistics systems: Logistics can be defined as the transport of material within manufacturing systems and supply networks. Logistics is a kind of manipulation of an object by humans, tools, machines, automation systems, and robots (or combinations of those), positioning and orientating objects to be transported or processed in a three-dimensional space. However, logistics operations do not change or transform the material directly. Logistics systems can be characterized according to various scales, such as assembly system scale, factory internal scale (*factory internal logistics*), factory external scale (*factory external logistics*).

Manipulator: In this book series, a manipulator refers to a system of multiple links and joints that performs a kinematic motion. Depending on the ratio of autonomy and intelligence, manipulators can be machines, automated systems, or robots.

Manufacturing: In this book, manufacturing refers to systems that produce products. Manufacturing integrates *production* (parts or low-level component production) and *assembly* processes.

Manufacturing lead time: Time necessary to complete a product within a given manufacturing system, factory, or factory network.

Mass customization (MC): MC strategies combine advantages of *workshop-like* and *production-line*–like manufacturing, and thus product differentiation–related competitive advantages, with mass-production–like efficiency. On the product side, MC demands that a product combines customized and standardized elements, for example, through *modularity*, *platform strategies*, and *frame and infill* strategies to be able to efficiently produce it in an industrialized manner. On the manufacturing side, MC demands highly flexible machines, automation systems, or robot technology that removes the need for human labour in the customization process.

Material handling, sorting, and processing yard (MHSPY): Subsystem of an *automated/robotic on-site factory*; often related to the *ground factory*. An MHSPY can be a covered environment and/or can be equipped with *overhead manipulators* and allows the simplification or automation of the picking up of *components* from delivering *factory external logistics* in a *just in time* and *just in sequence* manner. An MHSPY can also be used to transform *parts* and low-level *components* into higher-level components on-site. In *automated/robotic on-site factories* used to deconstruct buildings, MHSPYs can be used to transform higher-level *components* into lower-level *components* and *parts*.

Means of production: Means of production can be classified into human resources, equipment, and material to be transformed.

Modular flexibility: When the change of a product or the variation of a product is so intense that the *inbuilt flexibility* of a manufacturing system, a machine, or an *end-effector* cannot cope with it, a rearrangement or extension of the manufacturing system on the basis of *modularity* becomes necessary. *Modularity* can be generic (predefined process or system modules) or unforeseen (use/design of completely new modules, new configurations).

Modularity: Modularity refers to the decomposition of a structure or system into rather independent subentities. It can cover the functional realm as well as the physical realm. If structures or systems are nearly impossible to decompose, on both functional and physical levels, the artefacts are referred to as "integral." If systems can clearly be decomposed, on both functional and physical levels, artefacts are referred to as "modular." Clear modularity is, in construction practice, still a rare phenomenon, and conventional buildings show basic characteristics of integral product structures. *Automated/robotic on-site factories*, however, require strict modularity.

Module: In this book, in a hierarchical modular structure, modules represent elements on a hierarchical level above high-level *components*. *Parts*, *components*, and modules can be assembled into *units*, which are ranked higher than modules.

nth, $n - 1$, $n - 2$, $n - X$ floors: Inside the main factory (e.g., a *sky factory*) of *automated/robotic on-site factories*, work (component installation, welding, interior finishing, etc.) is done in parallel on several floors (n-floors). The nth floor represents the uppermost floor in which work takes place in parallel and the $n - X$ floors represent the floors below this floor in which work takes place in parallel.

OEM-like integration structure: Value systems or *parts/components* integration structures that do not fully follow the *OEM model* but show characteristics of it.

OEM model: An *original equipment manufacturer* relies on suppliers, which, according to their rank in the supply chain, are called Tier-n suppliers. The model explains the general *flow of material* as well as the flow of information during development of the product and its subcomponents.

Off-site manufacturing (OFM): *Components* or complete products are manufactured in a *structured environment* distant from the final location where they are finally used. *Components* or complete products can be packed and shipped or are mobile (e.g., car, aircraft).

One-piece-flow (OPF): OPF refers to a highly systemized and *production-line–*based manufacturing system in which each *component* or product assembled can be different.

On-site manufacturing (ONM): Products such as buildings, towers, and bridges have to be produced on site by ONM systems at the location at which they are to be finally used as they cannot be moved or shipped as an entity.

Open sky factory (OSF): *Sky factory* covering and protecting the workspace in *automated/robotic on-site factories*, and, in contrast to *closed sky factories* allows only the installation of a partly *structured environment* that at least (compared to conventional construction) minimizes the influence of parameters that cannot be 100% specified/foreseen (e.g., rain, wind).

Original equipment manufacturer (OEM): Integrates and assembles *components* and subsystems coming from sub-factories and suppliers to the final product within the *OEM model*. Companies or entities in the value chain that do not fully follow the *OEM model* but show characteristics of it are also referred to as *final integrators*.

Overhead manipulators (OMs): OMs operate within off- or on-site *structured environments* and in *automated/robotic on-site factories* are often the central elements of the *horizontal delivery system*. On the one hand, OMs (e.g., gantry-type OMs) allow the precise manipulation of components of extreme weights and at high speed, which cannot, for example, be accomplished by conventional industrial robots such as anthropomorphic manipulators. On the other hand, OMs require a simplification of the assembly process by *robot-oriented design*, as their workspace and their ability to conduct complex positioning and orientation tasks are limited.

Part: In this book, in a hierarchical modular structure, parts represent elements on a hierarchical level below *components*.

Performance multiplication effect (PME): Once significant productivity increases in an industry can be achieved (i.e., by switching from crafts-based to machine-based manufacturing), an upward spiral starts: high productivity can become a driver of the financing elements for innovations related to even better machines, processes, and products and thus even higher productivity. This phenomenon was/can be observed in many non-construction industries (e.g., textile industry, automotive industry, shipbuilding) and is in this book series referred to as the PME.

Platform strategy: A platform is a basic framework; a set of standards, procedures, or parts; or a basic structure that contains core functions of a product. A platform allows for the highly efficient production of customized products, as it allows for the platform to be mass-produced and to wear individual modules on top of it, which can be customized or personalized.

Positioning and orientation: For unrestricted positioning of an object within a defined space, or within x, y, and z coordinates, at least 3 *degrees of freedom* are necessary (also referred to as forward/back, left/right, up/down). For unrestricted orientation of an object around a *tool centre point*, at least 3 *degrees of freedom* are necessary (also referred to as yaw, pitch, and roll).

Production: In this book, production refers to the generation of basic *parts* or low-level *components*. It includes transformation of raw material into *parts*. *Downstream* processes dealing with the joining of elements generated within production are referred to as *assembly*. *Manufacturing* includes production and *assembly* processes.

Production line–like organization: In a production line–like organization, the flow of material between individual workstations is fixed; a material transport system links the stations and the *cycle times* of the individual workstations are synchronized.

Productivity: Productivity = Output (quantity)/Input (quantity). Productivity quantitatively expresses an input-to-output ratio, with a focus on a single (input) means of production or a single (input) factor of production. Productivity indices concerning the type of factor are, for example, work, capital, material, resource, and machine productivity.

Pulling production: Refers to a production system in which products are manufactured only on the basis of actual demand or orders. *Parts*, *components*, and products required are pulled from *upstream*, according to the actual demand. It might refer to the whole manufacturing system, as well as to individual workstations or groups of workstations. Examples: *Toyota Production System*, Sekisui Heim, Toyota Home.

Pushing production: Refers to the continuous production of elements/products in a certain fixed amount based on predictions or assumptions. Without taking into consideration the actual demand in *downstream* process steps, *parts*, *components*, and products are pushed through individual stations. It might refer to the whole manufacturing system or to individual workstations or groups of workstations. Example: Henry Ford's mass production.

Radio frequency identification tag (RFID): RFID tags are inexpensive tags that can be attached to *components*, *modules*, *units*, or products. RFID readers can be integrated into floors or placed over gates and can then identify the object passing by. They can be distinguished between simple low-cost passive tags and more complex active tags. Advanced readers can read multiple tags at once.

Real-time economy (RTE): Macroeconomic view of the impact of the multitude of changes our economy, manufacturing technology, and the relation between customers and businesses undergo. It targets the fulfilment of customer demands and requests in near real time. Products and services are processed within a few hours and delivered within a few days.

Real-time monitoring and management system (RTMMS): Data from sensor systems, as well as from the servomotors/encoders of the *vertical delivery system* (VDS) and *horizontal delivery system* (HDS), along with information obtained from cameras monitoring all activities (including human activities) in *automated/robotic on-site factories* are used to create a real-time representation of equipment activity and of construction progress. Furthermore, barcode systems often allow the representation and optimization of the *flow of material*, allowing equipment (such as VDS, HDS, or OM) to identify the *component* being processed. In most cases, real-time monitoring and management is done in a fully computerized on-site control centre. An RTMMS simplifies progress and quality control and reduces management complexity.

Re-customization: Remanufacturing strategy that allows a building to be disassembled and for major *components* or *units* to be refurbished and equipped with new *parts* or *modules* on the basis of mechanized or automated manufacturing systems to meet changed or new (individual) customer demands.

Robot-oriented design (ROD): ROD is concerned with the co-adaptation of construction products and automated or robotic technology, so that the use of such technology becomes applicable, simpler, or more efficient. The concept of ROD was first introduced in 1988 in Japan by T. Bock and served later as the basis for automated construction and other robot-based applications.

Rotation: A term used to describe a kinematic structure. A revolute joint allows an element of a machine or manipulator to rotate around an axis and in a serial kinematic system adds 1 *degree of freedom* to the system.

Selective compliance articulated robot arm (SCARA): Developed by Yamanashi University in Japan in the 1970s. It combines two revolute and one prismatic joint so that all motion axes are parallel. This configuration and the thus enabled allocation of the actuators are advantageous for the stiffness, repeatability, and speed with which the robot can work. Owing to its simplicity, the SCARA is also a relatively cheap robot system. It laid the foundation for the efficient and cheap production, and thus the success of, for example, Sony's Walkman.

Single-task construction robot (STCR): STCRs are systems that support workers in executing one specific construction process or task (such as digging, concrete levelling, concrete smoothening, and painting) or take over the physical activity of human workers that would be necessary to perform one process or task.

Sky factory (SF): Structured environment (factory or factory-like) setup on the construction site as part of *automated/robotic on-site factories*. SFs cover the area where building *parts* and *components* are joined to the final product and rise vertically with the upper floor of a building through a *climbing system*. SFs can enclose and protect the work environment completely (*closed sky factory*) or only partly (*open sky factory*).

Slip forming technology: Moving or self-moving form that allows casting concrete structures such as columns, walls, or towers on site in a systemized manner on the construction site.

Stilts: In this book, stilts refer to elements of automated/robotic on-site factories. The *sky factories* of *automated/robotic on-site factories* often use stilts (made of steel) integrated within the *climbing system* to be rested on the building that they are manufacturing. Stilts can be lifted and lowered via the *climbing system*, thus allowing the *sky factory* to move on top of the building's steel column structure.

Structured environment (SE): In factories or factory-like environments, work tasks, workspaces, assembly directions, and many other parameters (e.g., climate, light, temperature) can be standardized and precisely controlled. The structuring of an environment creates the basis for the efficient use of machines, automation, and robot technology. The structuring of an environment includes the protection from uncontrollable factors such as wind, rain, sun, and non-standardized human work activity.

Superstructure: The concept of dividing a building into superstructure and substructure is an approach that introduces the concept of hierarchies to a building's structure and *components* and thus can serve as a basis for possible *modularity* in construction. Goldsmith introduced the idea of making the transmission of forces within high-rise buildings by the superstructure to a clearly visible architectural element in his thesis (1953). A superstructure can serve as a platform or frame that allows customization by further infill and is thus closely connected to *frame and infill* strategies.

Supply chain: The supply chain connects value-added steps and transformational processes across the border of individual factories or companies. Its aim is to interconnect all processes and workstations to complete a product informationally and physically, in order to create uninterrupted on-demand or in-stock *flow of material*.

Sustainability in manufacturing: Manufacturing systems can be designed to be efficient and to equally meet economic, environmental, and social demands. In this book, sustainability in manufacturing refers predominantly to the ability of a manufacturing system to reduce consumption of resources and the generation of waste.

Technology diffusion: Technology diffusion describes the step-by-step spread of a technology throughout industry or as, for example, computer technology throughout society. To simplify the adoption of novel technologies and increase their application scope over time, novel technologies have to be made less expensive and less complex and be split into individual modular elements. Technology diffusion therefore often is accompanied by a switch from centralized to rather decentralized applications of the novel technology.

Tier-*n* supplier: Suppliers are, according to their rank in the supply chain, called Tier-*n* suppliers. A Tier-1 supplier is a first-rank supplier that relies on *components* from a Tier-2 supplier; a Tier-2 supplier relies on *components* from Tier-3 suppliers; and so on.

Tool centre point (TCP): The *end-effector* is a tool that is carried by the kinematic system. For each end-effector, a tool TCP is defined as the reference point for kinematic calculations.

Toyota Production System (TPS): The TPS is a logical and consequent advancement of the concept of mass production to a more flexible and adaptive form of demand-oriented manufacturing, developed by Toyota between the 1960s and 1970s. From the 1980s, TPS principles gained worldwide recognition and today they form the conceptual basis for manufacturing systems around the world. Concepts such as *just in time*, Kaizen, *Kanban*, *pulling production*, failure-free production, and *one-piece flow* have their origins in the TPS.

Transformational process: Any organization and its manufacturing system transform inputs (information, material) into outputs (products, services). The transformation is performed by the organization's structural setting and its means of production, resulting in a specific combination and interaction of workers, machines, material, and information.

Translation: A term used to describe a kinematic structure. A prismatic joint allows an element of a machine or manipulator to move in a given trajectory along an axis and, in a serial kinematic system, adds 1 *degree of freedom* to the system.

Tunnel boring machine (TBM): TBMs mechanize and automate repetitive processes in tunnel construction. TBMs are self-moving underground factories that more or less automatically perform excavation, removal of excavated material, and supply and positioning of precast concrete segments. TBMs are equal in many ways to automated construction sites (and the production of a building's main structure by *automated/robotic on-site factories*).

24/7-mode: The operation of a factory, processes, or equipment without major interruption 24 hours a day, 7 days a week. Requires the work environment to be structured (*structured environments*) and protected from influencing factors such as weather and the day/night switchover.

Unit: In this book, units refer to high-level building blocks. *Parts*, *components*, and *modules* can be assembled into units, which are higher ranked than *modules*. Units are completely finished and large three-dimensional building sections, manufactured off site.

Unit method: Sekisui Heim, Toyota Home, and Misawa Homes (Hybrid) break down a building into three-dimensional *units*. Those *units* are realized on the basis of a three-dimensional steel space frame, which, on the one hand is the bearing (steel) structure of the building, and on the other hand can be placed on a production line where it can be almost fully equipped with technical installations, finishing, kitchens, bathrooms (plumbing *units*), and appliances.

Upstream/downstream: Refer in this book to processes or activities in a value chain or manufacturing chain that are conducted before (upstream) or after (downstream) a certain point.

Urban mining: Refers to strategies that allow the city and especially its building stock to be a "mine" for resources, *parts*, and *components*. Systemized deconstruction of buildings under controlled and structured conditions, as in *automated/robotic on-site factories*, are enablers of urban mining.

Vertical delivery system (VDS): System that transports *parts/components* on the construction site from the ground (e.g., material handling yard) to the floor level where the components are to be assembled. VDSs play an important role in most *automated/robotic on-site factories*.

Workbench-like organization: The product stays at a fixed station in the factory where it is produced or assembled manually or automatically through the use of various tools. The means of production are organized around this one station.

Workshop-like organization: In a workshop-like organization, the product and/or its components flow between workstations. The sequence is not fixed and the times products stay at a certain workstation vary with the product.

"Zero" waste factory: A factory that minimizes resource input and waste output and allows (almost) all generated waste to be recycled.

1 Integrated Automated/Robotic On-site Factories

In this volume all worldwide conducted approaches following an on-site factory approach were analysed. Thirty different systems were identified, resulting in an application of automated/robotic on-site factory technology about 60 times. The analysis was for each system split into a more technical part and a part that focuses on indicators related to productivity, efficiency, and economic performance. All systems were analysed systematically and based on the same framework. On the basis of the analysis, a categorization system was developed and 13 categories were set up (10 categories for construction and 3 for deconstruction).

As discussed in **Volume 3**, one of the main ideas for setting up automated on-site factories was to integrate stand-alone or single-task construction robot (STCR) technology in structured on-site environments into networked machine systems and thus to improve through interlinked machine activities the organization, integration, and material flow on the construction site (apart from the possibility to off-site manufacture components discussed in **Volume 2**). The analysis clarifies for which building typologies automated/robotic on-site factories are an applicable approach and how and to which extent each of those systems is technologically flexible to be able to manufacture a variety of different buildings (products) on the basis of industrialized, automated, and flow-line–like stable factory processes on the construction site. Furthermore, it should be clarified whether, in contrast to the STCR approach (see **Volume 3**), the approach of setting up automated/robotic on-site factories is capable of achieving a performance multiplication (e.g., by 10-fold as in tunneling or automotive industry; for further details see **Volume 1**), which usually accompanies the switch from arts and crafts–based manufacturing to machine-based manufacturing.

In this volume, first, frameworks for the technical analysis and the efficiency analysis classified into various fields, analysis subjects, and indicators are set up. Second, 30 systems are analysed according to the technical analysis framework and classified into two main categories (construction: 24, deconstruction: 6) and a total of 13 subcategories. Ten of those systems are also analysed with regard to analysis subjects and indicators that determine or influence productivity, efficiency, and overall economic performance. An *Analysis and Categorization Matrix* (see Figure 2.35) gives an overview over systems, categories, analysed analysis subjects and indicators, and available data. Finally, the findings are summarized (see Chapter 4). The analysis claims to include all approaches to automated/robotic on-site factories that were conducted so far.

1.1 Framework for Technical and Efficiency Analysis

In this volume, the technical aspects of integrated automated construction sites (including their composition and configuration), as well as their resulting efficiency, are analysed. Automated construction sites represent on-site manufacturing (ONM) environments (fixed type or moving type) that conduct a final assembly of low-, medium-, and high-level modular components rather than a conventional construction process. The installation of a factory on-site structures the work environment and allows the application of automation and robot technology. Furthermore, the processing of value-added components designed according to robot-oriented design (ROD) strategies reduces on-site complexity. The modularization of both building products and manufacturing systems allows for flexibility and customisation of the products (buildings) to a certain degree. The technical analysis follows the identification of concepts and strategies relevant to the fields of multilevel modularity (see **Volume 1, Section 4.2**), manufacturing technologies and strategies (see **Volume 1, Section 4.3**), and automation and robot technology (see **Volume 1, Section 4.4**). An overview of the framework developed for the technical and configuration analysis is presented in Table 1.1. The data used in this analysis were acquired from various sources such as project descriptions by the companies, publications by companies and their R&D staff, publications by researchers who had analysed systems, expert interviews with developers and company staff, and own site visits. Furthermore, as a basis for the analysis, documentary material in the form of plans, project descriptions, and an own picture archive documenting the application of nearly all systems were used.

As far as efficiency and productivity are concerned, the analysis framework is based on the assumption that technical and economic efficiencies are generated by both an efficient combination of input factors and the set-up of a high-value product with a low defect rate. Individual performance indicators, such as work productivity, material efficiency, physical strain, health and safety, construction quality (related to the defect rate), and integration along the value chain have been identified as the most influential construction performance indicators (see **Volume 1**). In addition, as outlined in **Volume 1,** the construction industry is highly labour intensive, with decreasing labour productivity, a high rate of construction defects, a high rate of fatal and nonfatal injuries, and a relatively high amount of material input compared to the output value. This correlates with the low investment and R&D spending rate and the low capital stock, indicating that the value and quality of the existing manufacturing equipment, process technologies, and skilled workforce are low. However, integrated automated construction sites would require, for example, as in the automotive industry, a high investment and R&D spending rate, a high level of capital stock, and in that context of course demand for considerable improvements in the aforementioned performance indicators. The efficiency analysis for each system attempts to analyse whether integrated automated site technology has the potential to deliver the demanded efficiency improvements. The technical configuration of the systems and their efficiency are closely related to each other. In most cases, positive

Table 1.1. *Framework for the analysis of technical aspects and system configuration*

Field of analysis	Analysis subjects and indicators
Evolution scheme	Location of sky factory and ground factory Working direction General workflow
Elevation	Detailed vertically organized workflow Parallel work on various levels Configuration of main and sub-factories Analysis of the component installation process
Ground plan (main and sub-factory)	Detailed horizontally organized workflow Configuration of main and sub-factories
Subsystems	SF (covered, working platforms, closed sky factory, open sky factory) Vertical logistics (in particular vertical delivery systems) Horizontal logistics (in particular horizontal delivery systems) Manipulators (in particular overhead manipulators) Climbing mechanisms (in particular climbing systems) Assembly simulation and progress control tools (real-time monitoring and management system and material handling, sorting, and processing yard)
End-effectors	Types of end-effectors Relation of end-effectors and components/materials Modularity Possibility of tool changes
System variations	Realized system variations Possible system variations Inbuilt flexibility Modular flexibility
Robot-oriented design (ROD)	ROD on a component level ROD on a building level ROD on an urban level

or negative efficiency performance can be directly correlated to the general set-up, configuration, use of subsystems/end-effectors, and the deployment of ROD.

The analysis framework for the systems' efficiency was synchronized with the currently available data sets. Analysis subjects that companies deploying the systems did not analyse or make available (e.g., detailed data on investment or defects/errors, or injuries related to the application of the new technology) were not considered in the framework. The *Analysis and Categorization Matrix* (see Figure 2.35) shows which data were made available for each system. Obviously, companies that deploy their systems more often than others (Obayashi, Shimizu) also generate or are interested in generating more data sets. Table 1.2 outlines the analysis framework and shows which analysis subjects and indicators were considered as relevant for the efficiency analysis.

All of the data presented have to be considered from the perspective that all systems were still in an experimentation, prototype or test phase. With the development of such technologies, Japanese companies have aimed at long-term efficiency

Table 1.2. *Efficiency analysis framework*

Thematic field	Analysis subjects	Indicators
Erection speed	Project schedule	Time necessary to set up a sky factory Operation period Time necessary for dismantling of a sky factory Floor production rate per month
	Floor erection cycle	Time and work steps necessary to complete a standard floor Parallel processing on several floors Equipment (e.g., overhead manipulator) operation sequence
Configuration	Technical data speed of equipment	Operational speed of horizontal delivery system Operational speed of climbing system Operational speed of vertical delivery system
	Experiment's degree of automation/system configuration/worker teams	Rate of automation – is installation operation remote controlled, partly automated, or fully automated? Could companies apply system in various configurations (flexibility) Influence of varying numbers of workers, e.g., on productivity
Productivity	Productivity workers/time (including comparisons with conventional construction or other systems)	Man-hours required for completion of floor Number of construction workers required for a specific task field Total number of workers Comparison with conventionally constructed buildings
	Learning effects	Reduction of time needed to install components with the novel site technology/equipment Reduction of time needed for welding Reduction of time required for factory internal logistics
Resource efficiency	Material and resource efficiency	Reduction of required input material Reduction of construction waste generated
Quality, health and safety	Product or process monitoring (real-time progress, decibel, etc., control room)	Real-time supervision of operations Real-time progress control, real-time monitoring and management system Simulation of optimized operation
	Safety	Influence of environment on safety
	Physical strain (heart rate, etc.)	Influence of environment on physical strain
	Weather influence	Influence of weather on: • Operation/task execution • Productivity • Quality
Usability studies	Evaluation of usability of on-site factory and equipment by workers/operators	Influence on work tasks Influence on motivation Influence on user acceptance and emotions

and at building up knowledge step by step, and thus made a number of compromises concerning short-term efficiency. Obayashi and Shimizu, for example, each of which deployed their systems in a multitude of construction projects (ABCS: applied six times; Big Canopy: applied six times; SMART: applied six times), introduced improvements in each new project and experimented with the configuration of the robotic crane systems or with the automation degree and the number of workers used (and thus the degree of work productivity). For example, during the first projects using the SMART, Shimizu had the intention of training its workforce on the general use of the new technology (according to Japanese philosophy, knowledge about new technologies and tools has to be spread fast among the workforce by special training procedures), and thus replaced half the workforce with new workers from project to project to train as many of their workforce in using the new technology as quickly as possible. Considering the fact that the learning effects within projects were enormous (see later in the efficiency analysis of the SMART system in Chapter 2), it can be assumed that this procedure influenced efficiency considerably, as it mitigated the impact of these learning effects across projects. Furthermore, most systems were still in a developmental phase, and did not fully utilize the capacity of their subsystems at any time during individual projects. Shimizu, for example, had up to 24 robotic trolleys (that can potentially be operated in parallel) available in a fully deployed SMART, but operated some projects with, for example, only 10 of them in an active mode.

Analysis and Categorization Matrix

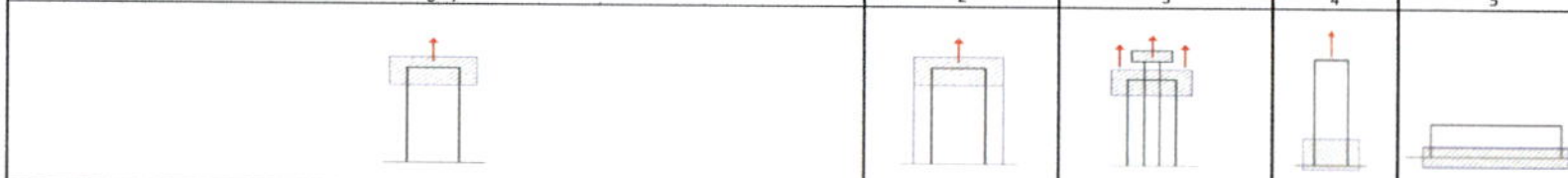

			1	2	3	4	5	6	7	8	9	10	11	12	13	14	15
			Construction by Automated/ Robotic On-site Factory														
			Sky Factory (moving upwards) supported by building								Sky Factory (moving upwards) on stilts (prolonging)		Sky Factory pulled up along core (main factory and core factory moving upwards)		Ground Factory (staying fixed place, vertically oriented building) and building push-up	Ground factory (staying fixed place, horizontally oriented building) and building push-up	
			Category 1								2		3		4	5	
			Obayashi	Fujita	Goyo	Maeda	Takenaka	Toda	Shimizu	BAM	Obayashi	Shimizu	Taisei	Korean Cons.	Kajima	Skanska	Sekisui House
			Automated Building Construction System (ABCS)	Akatuki 21	Future Automated Construction Efficient System (FACES)	Mast Climbing Construction System (MCCS)	Roof Push-up	Roof-Robo	Shimizu Manufacturing System by Advanced Robot Technology (SMART)	System Netherlands	Big Canopy	SMART Light	Totally Mechanised Construction System for High-rise Buildings (T-Up)	Robotic and Crane based Automatic Construction System (RCACS)	Automatic Up-Rising Construction by Advanced Technique (AMURAD)	System Skanska	J-UP
Technical Analysis	Factory Layout (Evolution Scheme, Elevation, Ground Plan)	Location of main factory	On top	On top	On top	On top	On top	On top	On top	On top	On top	On top	On top	On top	On ground	On ground	On ground
		Working direction	Vertically upwards	Vertically upwards	Vertically upwards	Vertically upwards	Vertically upwards	Vertically upwards	Vertically upwards	Vertically upwards	Vertical upwards	Vertical upwards	Vertical upwards	Vertical upwards	Push up vertical	Push up vertical	Push up vertical
		Building Material Main Structure	Seel	Seel	Seel	Seel	Seel	Seel	Seel	Concrete	Concrete	Concrete	Steel	Steel	Concrete	Concrete	Steel
	Sub-systems	Main Factory Cover Structure	x	x	x	x	x	x	x	x	x	x	x	x	x	x	x
		Vertical Delivery System		x			x	x			x	x		x	x		
		Horizontal Delivery System		x		x	x	x		x	x	x		x	x	x	
		Vertical/Horizontal Delivery System	x		x		x		x	x		x	x				
		Number of Overhead Manipulators	3	3	2	2	2	2	Up to 25	2	Up to 3	Min. 3	2	2	3	1	
		Climbing System (or push-up/sliding)	x	x	x	x	x	x	x	x	x	x	x	x	x	x	x
		Dedicated Facade Installation System	x														
		Sub-factory (storage, component preparation)		x				x	x			x	x		x	x	x
		Column Welding Subsystem	x		x	x			x								
		Beam Welding Subsystem	x														
		Logistics Robots for Interior Finishing	x						x				x		x		
		Other Robots for Interior Finishing					x	x	x				x		x		
		Material Management System	x	x	x	x			x		x		x	x	x	x	x
		Real Time Monitoring System	x	x	x	x	x	x	x		x		x	x	x		
		Component Alignment System	x		x	x		x					x				
		Construction Process Simulator	x		x				x		x			x			
		Number of Sub-systems	10	12	7	12	9	10	8	5	7	6	13	6	9	8	6
	End-effectors	Numer of End-effectors	6	4	7	1	1	1	1	2	2	1	4	>1	7	1	1
		In-built Flexibility (+, ++, +++)	+	+	+	+++	+++	+++	+++	+++	++	+++	++	+	++	+++	+++
		Modular Flexibility (+, ++, +++)	+++	+++	+++	+	+	+	+	+	++	+	++	+++	+++	+	+
	System Variations	Number of applied and planned Variations	4	2	2	2	1	1	6	1	3	2	6	1	3	1	2
	Robot Oriented Design (ROD)	ROD on component level	x	x	x	x	x	x	x	x	x		x	x	x	x	
		ROD on building level						x					x	x		x	
		ROD on urban level															
Efficiency Analysis	Erection Speed	Total on-site construction time necessitated	6-24 month		15 month						2 years		20 month		10 month		2-4 month
		Time necessitated for system set up	1 month	2 month	2-5 month	1-2 month	1-2 month	1-2 month	1-2 month	2-3 month	1 month		6 weeks		6 weeks		1 day
		Time necessitated for system dismantling	2-3 weeks	2 weeks	1 month	2-3 weeks	2-3 weeks	2-3 weeks	2-3 weeks	1-2 month	1 month		2 weeks		2 weeks		1 day
		Floor Erection Cycle	5/6 days	9 days	6 days	6-8 days	5,5 days	6-8 days	5/6 days	14 days	7 days	about 10 days	3 days		6 days		
		Number of Floors produced per month	4	2-3	2,5	2-3	2-3	2-3	about 3	1-2	2,7	2	2,4		2		1-2
		Start of operation form floor X	3	3	0	2	3	4	1		3	0	5-6	1	2	1	1
		Parallel work levels	>4	3	>4										6		
		Levels enclosed by Sky Factory	3	3	4										3		
	Configuration	Speed of eqipment	40 m/min	20-25 m/min	20 m/min	20-25 m/min	20-25 m/min	20-25 m/min	20-40 m/min		40 m/min	20-40 m/min			10 m/min	10-20 m/min	
		Automation ratio	Automatic/ remote	Automatic/ remote	Automatic/ remote (automatic mode was slower)	Automatic/ remote	Automatic/ remote	Automatic/ remote	Automatic/ remote	Remote controlled	Full automation and remor operation possible (remote 10% faster)	Remote	Automatic/ remote	Automatic/ remote	Automatic/ remote	Automatic/ remote	Remote
	Productivity	Task specific work productivity			Improvement of work productivity by the two to six fold				Manhours reduced by 50%, number of workers reduced by 30%								
		Total labor productivity							Time necessary to construct the building reducecd by about 15%,		Up to 70% less workers		Manhours 68% reduction, construction time: 20% reduction		Improvement of labor productivity of up to 50%, reduction of construction time up to 30%		
		Learning effects	Component installation 10%, welding 20%						20 % for column positioning		Up to 20%				40-50% for standard floor		
	Ressource Efficiency	Measurements to reduce consumption/ waste							70% waste reduction		50% waste reduction						
	Quality, Helath and Safety	Multible working platforms	x	x	x	x	x		x	x					x		
		Protection in terms of	Sun, wind, temperature, rain, sound	Sun, wind, temperature, rain, sound	Sun, wind, temperature, rain, sound	Sun, wind, temperature, rain, sound	Sun, wind, temperature, rain, sound	Sun, wind, temperature, rain, sound	Sun, wind, temperature, rain, sound	Sun, wind, temperature, rain, sound	Sun, rain	Sun, rain	Sun, wind, temperature, rain, sound	Sun, wind, temperature, rain, sound	Sun, wind, temperature, rain, sound	Sun, wind, temperature, rain, sound	Sun, rain
	Usability	Usability aspects	Positive influence of protected environment		Attract skilled workers by technology				Attract skilled workers by technology		Measurable decrease of heart-rate of workers						

									De-construction by Automated/ Robotic On-site Factory					
Off- and on-site combined factory (both off and on-site factory fixed place)	Self-supported Ground Factory (horizontally moving)		Sky Factory (moving upwards) for simple tower manufacturing		Centralized Sky Factory (moving upwards) in combination with conventional construction		Decentralized Sky Factory (moving upwards) in combination with conventional construction		Closed Sky Factory supported by building (moving downwards)		Open Sky Factory supported by building (moving downwards)			Ground Factory (fixed place) and building lift-down
6	7		8		9		10		11		12			13
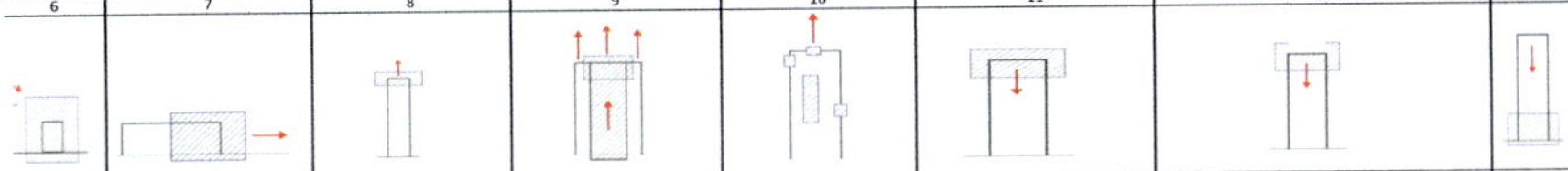														
16	17	18	19	20	21	22	23	24	25	26	27	28	29	30
NCC	Summerfield	Neufert	Taisei	Shimizu	Obayashi	Shimizu	Obayashi	Shimizu	Takenaka	Taisei	Nishimatsu	Shimizu	Obayashi	Kajima

NCC Komplett	Bauschiffverfahren	Hausbaumaschine	TS-Up	Tower SMART	Hybrid-ABCS	Hybrid-SMART	Loose Deployment of ABCS-Subsystems (e.g. Tokyo Skytree)	Loose Deployment of SMART-Subsystems (e.g. Mode Gakuen Cocoon Tower)	Hat Down	Taisei Ecological Reproduction System (TECOREP)	Move Hat	Reverse Construction Method (RCM)	Quakeproof, Quiet, Quick and Block-by-Block Building Disassembly (Cube Cut Method)	Cut and Take Down Method "Daruma Otoshi" (DARUMA)
Covers building	Covers building	Covers building	On top	On top	On top	On top	Distributed	Distributed	On top	On top	On top	On top	On top	On ground
Vertical/ Horizontal	Horizontal	Horizontal	Vertical	Vertical	Vertical + around	Vertical + around	Vertical	Vertical	Vertical	Vertical	Vertical	Vertical	Vertical	Vertical
Concrete	Concrete	Concrete	Steel	Steel	Steel	Steel	Steel	Steel	Steel/ concrete	Steel/ concrete	Steel/ concrete	Steel/ concrete	Concrete	Concrete
x	x	x	x	x	x	x		x	x	x	x	x	x	x
	x	x		x			x	x						x
	x	x		x					x	x				x
x			x		x	x	x	x	x		x	x	x	
2	2	6-20	-	1	3	Up to 25	-	-	>3	>2	>2	-	-	-
	x	x	x	x	x	x		x	x	x	x	x	x	x
					x			x			x			
x						x		x	x	x	x			x
					x	x	x		x	x				x
					x		x							
					x	x								x
						x							x	
x					x	x	x	x	x	x				x
					x	x			x	x				x
					x		x							
					x	x								
6	4	5			10	9	7	6	6	6	5	5	5	7
1	3	2	1	1	6	1	1	2	1	1	1	-	-	
+++	+	+	+	+	+	+++	+++	++						
+	+	+	+	+	+++	+	+	++						
2	3	1	1	1	3	2	>4	>4	1	2	2	2	2	2
x		x	x	x	x	x		x						
	x	x	x	x	x	x								
	x	x												
3-6 month					6-24 month				6 month	1 year	Same as conv. construction	Same as conv. Construction	Same as conv. construction	6 month
2 weeks					1 month	1-2 month								
1 week					2-3 weeks	2-3 weeks								
					5/6 days	5/6 days			6-10 days	6-10 days	6-10 days	6-10 days	6-10 days	6-10 days
2-4					4	about 3								
-1	0	0	2	2	3	1	0	0	1	1	0	0	0	-
					>4									
					3									
20-40 m/min				20-40 m/min	40 m/min	20-40 m/min	40m/min	20-40 m/min						
Remote	-	-	Remote	Remote	Automatic/ remote	Automatic/ remote	Automatic / remote	Automatic / remote	Remote	Remote	Remote	Remote	Remote	Remote
						Manhours reduced by 50%, number of workers reduced by 30%								
NCC aims at enhanced productivity (e.g. 4-6 buildings per worker per year			Shortening of construction time by 20%	Shortening of construction time by 20%		Time necessary to construct the building reduced by about 15%,								
					Component Installation 10%, Welding 20%	20% for column positioning								
						70% waste reduction			> 90% recycling rate					96 recycling rate
		x	x		x			x	x	x	x	x		
Sun, wind, temperature, rain, sound	Sun, rain	Sun, wind, temperature, rain, sound	Wind	Sun, wind, temperature, rain, sound	Sun, wind, temperature, rain, sound	Sun, wind, temperature, rain, sound		Sun, wind, sound	Sun, wind, temperature, rain, sound, dust	Sun, wind, temperature, rain, sound, dust	Wind, sound, dust	Wind, sound, dust	Wind, sound, dust	Sun, wind, temperature, rain, sound, dust
Heated environment enhanced work speed and motivation					Positive influence of closed environment of conetration	attract skilled workers by technology			Safe deconstruction environment	Safe deconstruction environment				

2 Analysis and Categorization: Construction

Integrated automated construction sites can be categorized according to various features or characteristics, such as general working directions, logistics strategies, climbing mechanisms, or configurations of the site factories. It is also possible to characterize them according to manufacturing views (organizational view, product variation view, order-oriented view, or location-oriented view; for further information, see **Volume 1**). Furthermore, a categorization could also be based on the main building materials processed (steel-based components, concrete-based components). However, the general working direction and thus the location and workflow orientation of the factory and the location of the majority of work activities play a major role in manufacturing of buildings and determine logistics strategies and factory configurations (see **Volume 1**). As buildings are complex and large products (similar to aircrafts or tunnels; see **Volume 1**) that require a final assembly on the fixed, final site (see **Volume 1**), the orientation of the building and thus the location and working direction on-site determine the general organizational setting and thus the logistics strategy, climbing system (CS), and factory configuration. In this chapter a combined location and working-direction–oriented view is used as the basis for categorization.

A location and working-direction–oriented view on manufacturing considers the location or environment in which the product is manufactured as well as its geometrical characteristics. Usually, a product can be manufactured off-site in a factory some distance from the final location to which the finished product is finally shipped and used, as the product in its final state is still a mobile entity (a kind of large module) for which a transport infrastructure exists. Most complex products, such as ships, automobiles, aircraft, and most consumer products can be produced that way. It is no problem to pack and ship them.

However, products such as buildings, towers, bridges, and so forth have to be produced on-site at the location at which they will finally be used and they simply cannot be moved or shipped as a complete entity. High-level component manufacturers or unit manufacturers such as Sekisui Heim circumvent the need for on-site production by splitting up a building into three-dimensional high-level modules that are then produced and finished in the factory so that only minor work (in the case of Sekisui Heim 15–20%; presented in detail in **Volume 2**) has to be done on the construction site. The automated construction system AMURAD, for example, follows a different strategy and produces the building by an automated/robotic assembly system on the final site and at a fixed place (on-site manufacturing [ONM], fixed-site type; see also Section 2.4.1) by using not high-level but low- to medium-level precast and prefabricated components. The prefabricated components are delivered just in time and just in sequence from an informationally integrated precast plant. Similarly, the erection of a tunnel by a tunnel boring machine (TBM; see also **Volume 2**) with a mechanized or automated component/segment assembly system is an on-site production directly linked to off-site manufacturing (OFM) in the form of the delivery of prefabricated concrete segments. Unlike AMURAD, however, a "TBM factory" is moving and contains element of a production line on-site (moving type). The AMURAD on-site factory system stays during operation more or less at a fixed location (fixed-site type) and "extrudes" the building. In the case of ABCS the factory moves upwards and, in the case of Sommerfeld's factory system, it moves horizontally to produce the building.

Civil engineering construction sites in tunnelling are also referred to as "line construction sites". The strict organization of all work activities along an axis or line simplifies systemic workflow organization and (finally) permits mechanization or automation by TBMs (for further details, see **Volume 1**). However, building construction and the floor-by-floor direction that the erection of a building follows demand more complex logistics and work procedures. Whereas a TBM can erect the main structure of a tunnel with one erector placing the segment, the erection of a main structure of a building or floor (columns, beams, floor slabs etc.) requires both a better integration of vertical (along the main direction) or horizontal (along the sub-direction) active logistics and component positioning processes and thus higher flexibility with regard to kinematics and end-effectors due to higher variability.

In addition to the main factory (e.g., sky factory [SF]), many companies deployed a sub-factory (e.g., on ground), where parts and components delivered to the site are stored (intermediate or buffer storage), assembled into more complex components, or prepared for lifting to and processing by the main (sky) factory (e.g., Akatuki 21). The advantage of this approach is that parts and lower level components (can be stored in a compact way on the transport system) can be transported to the site more efficiently than complex and larger components (often demand that due to size and placement in templates a considerable amount of air is shipped). Although the logistics strategy, in terms of efficiency of the systems, might be a key element, it was not be considered as a basis for the categorization, as this strategy can basically be applied to any automated construction site and thus is not a unique characteristic. However, whenever companies have used sub-factories the approach was carefully analysed.

The analysis showed that some deconstruction systems working with downwards moving SFs also utilized the ground floors of a building as a sub-factory/ground factory (GF) where high-level components coming from the SF were processed into low-level components or mono-material for efficient transport to the recycling facilities. ONM systems share a need to be highly mobile, modular, easy to deploy, and dismountable as they are only temporarily installed. For example, Shimizu has developed a simulator for the SMART for optimizing the configuration, installation, and disassembly of the factory on a specific site and for a specific building.

Compared to OFM systems, so far ONM systems tend to have a lower efficiency and lower degrees of automation. A reason for this is that up to date usually a relatively long time is needed to set up the site factory and the complex and relatively heavy on-site automation and robot systems that were still in a developmental phase required slow and careful operation. However, current trends in automation and robot technology, such as modularity, plug-and-play, and lightweight design, have the potential to overcome these obstacles and lower the barrier to entry (for further discussion of the capability of automation and robot technology, see **Volume 1**).

Some companies also have tried to avoid the necessity for disassembly of the on-site factory (e.g., Obayashi, ABCS) by integrating the frame of the factory as a frame for the final floors into the building after completion. Taisei's deconstruction system tries to avoid both the assembly and the disassembly of the on-site factory by utilizing one of the top floors of the building as a factory ceiling and the basis for the installation of overhead manipulators (OMs).

2.1 Sky Factory (moving upwards) – Supported by Building

Systems in this category are based on the erection of an SF on top of the building to be constructed. The buildings are assembled from bottom to top. In the SF the building is assembled floorwise and the newly constructed floors are used as the supporting structure that the climbing mechanism of the SF uses to raise the factory upwards.

Three types of climbing mechanisms can be identified:

1. The factory climbs using stilts that use the vertical column structure as support (ABCS, Akatuki 21, FACES, MCCS, Roof Push-up, Roof-Robo).
2. The stilts carrying the factory are supported not by resting on top of individual columns but by using a bridging system and fixing them in between columns and on beams connecting the columns (SMART).
3. The factory rests on the building through girder framework bridges (System Netherlands).

Climbing mechanisms type 2 and type 3 have the advantage that the positioning and fixation of columns on top of other columns is not hindered by the climbing mechanism. However, climbing mechanism type 1 can be used as a fixture or jig, which helps to guide components to be positioned (e.g., columns) into place. Some systems are based on the idea of using a GF (assembly of low-level components to medium or high-level components; Akatuki 21, Roof-Robo) and others focus more on the direct pick-up of components from trucks. However, numerous systems are tried from application to application, and basically all systems in this category have the potential of using GF, direct pick-up, or a combination of both approaches. Systems in this category also vary considerably with regard to the type of main manipulators (automatic/robotic logistics and positioning systems) used.

While some companies build on a reduced number of high-capacity, large and heavy (but highly accurate) manipulators (e.g., FACES), others follow the strategy of distributing complexity and build on a multitude of smaller and faster manipulators that can be operated in parallel (e.g., SMART). A further characteristic of systems in this category is that the SFs completely seal the working environment from the outside and create weather-protected and soundproof environments. Accordingly, most systems in this category organize much of the vertical logistics within the building itself, using shafts as logistics channels. Thus, GF work, vertical logistics, and work in the SF are not affected by the weather. Furthermore, SFs encompass the building completely and also provide working platforms on the outside of the facade, which allows the finishing and sealing of the facade within the factory and as part of the floor erection cycles (for further information on floor erection cycles, see *Efficiency Analysis* of each system).

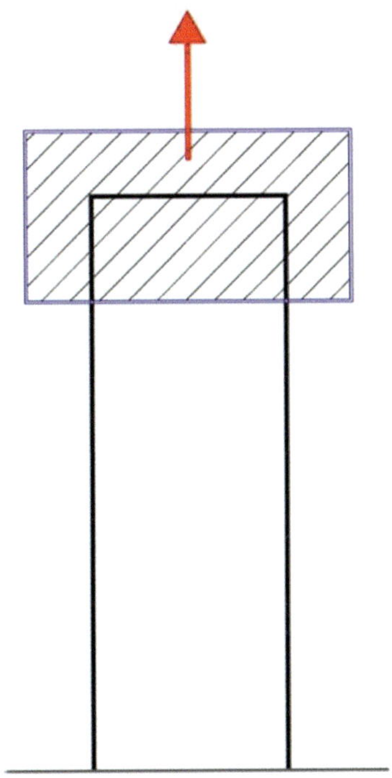

	System name	Company
1	Automated Building Construction System (ABCS)	Obayashi
2	Akatuki 21	Fujita
3	Future Automated Construction Efficient System (FACES)	Goyo/Penta Ocean
4	Mast Climbing Construction System (MCCS)	Maeda
5	Roof Push-up	Takenaka
6	Roof-Robo	Toda
7	Shimizu Manufacturing System by Advanced Robot Technology (SMART)	Shimizu
8	System Netherlands	Royal BAM

2.1.1 Automated Building Construction System (ABCS)

Company: Obayashi Corporation, Japan
Category: SF (moving upwards) – supported by building

Figure 2.1. ABCS SF automatically constructing a high-rise building floor by floor.

Analysis of Technology and Composition

Evolution Scheme ABCS concentrates work activity on a main factory that covers the top four to six floors to create a structured work environment (Figure 2.1). Within this work environment the steel main structure is erected, the facade is installed and sealed, and the interior is completely finished, so that when the factory moves upwards a completely finished building (or floor as the factory moves floor-by-floor) is left below. Usually, the factory is installed on top of the first two or three floors and starts operation from these floors onwards. The installation of the site factory at the beginning is time consuming (3–6 weeks); however, during full operation and after a learning phase, the system can fully complete up to two floors per week. To simplify the disassembly of the site factory, the steel frame structure of the factory is, in some projects, integrated as the load-bearing structure of the top floors of the building being manufactured.

Elevation The components are delivered to the factory by vertical delivery systems (VDSs, automated lifts). In the factory, the components are picked up from the lift in a just in time (JIT) and just in sequence (JIS) manner by other horizontally operating subsystems. Depending on the required capacity, the number of vertical lifts can be varied. In most cases so far, more than one lift has been installed to allow for a continuous vertical material supply (while one lift was going down to pick up the next component the other lift had already delivered the next one). The automated vertical lifts could be positioned either on the outside of the building (longitudinal side) or within the floor area in shafts that can be closed after building completion or used, for example, as elevator shafts. ABCS does not explicitly install a GF, but uses the ground floors for storage, delivery, and material sequencing purposes.

Ground Plan The horizontal delivery and positioning of components is done by gantry and overhead type manipulators (guided by a fixed rail system mounted to the roof structure of the factory) that can exchange their end-effectors. The manipulators can be used for picking up components from the vertical delivery system (VDS) and directly installing them or only for vertical transport and handing over to other manipulators. The installation of facade elements can be done either by the horizontally operating manipulator (in conjunction with a dedicated end-effector) or directly by an additional (and optional) VDS that is used only to deliver and install facade elements directly.

Subsystems The ABCS manufacturing the building on-site consists of a multitude of integrated subsystems:

1. *Factory roof structure.* Covers the workplace completely, covers the vertical logistics system, provides structure for mounting of rail system for operation of horizontal manipulators, provides covered and safe working platforms for facade work (Figures 2.8, 2.18, 2.19, 2.20, 2.21, 2.22, 2.35, 2.36, 2.37, 2.41, and 2.47).
2. *Climbing system.* The factory rests and ascends atop previously assembled steel columns and is fixed to hydraulically powered stilts that bear the weight of the roof structure and push it upwards floor by floor. The stilts are positioned over each column; only every second stilt is needed to bear the factory; thus it is possible for every second stilt to be lifted to enable the positioning of a new column by the manipulator under it (Figures 2.10, 2.11, 2.38, 2.39, 2.40, and 2.48). The CS is not used as a physically active fixture (as with FACES) but at least guides the installation process by providing reference points for the alignment and accuracy measurement system (AAMS).
3. *Vertical delivery system.* Automatic lifts of different capacities that deliver material up from the ground to the SF (Figures 2.6, 2.7, and 2.23). The lifts can be extended with the upwards moving SFs simply by placing new lift elements on top of the lift system.
4. *Horizontal delivery system.* Gantry-type OMs with exchangeable end-effectors (Figures 2.9, 2.12, 2.13, 2.14, 2.15, 2.17, 2.24, 2.25, and 2.28). Manipulators are rail guided – the gantry system is able to move (parallel translation) along two rails lengthwise, to move the end-effector perpendicularly, to rotate and to lower components. With these four degrees of freedom (DOFs) the manipulators are able to reach any point within the working space. In cooperation with the

end-effectors additional DOFs are gained for component positioning. Within each lengthwise rail system one manipulator moves. Depending on the width of the building, one to three rail systems (and thus manipulators) can be installed side by side.

5. *Facade element delivery and installation system (optional).* Under the factory roof, along the facade (outside of the floor area but along the building facade/edges), an additional rail-guided delivery system can be installed that vertically delivers and vertically/horizontally positions facade elements. The system disburdens the vertical and horizontal delivery systems (HDSs) mentioned previously and allows for further parallelization of manufacturing processes.
6. *Welding robot system for columns.* The welding robot systems for columns are mobile single-task construction robots (STCRs) that have been made an integral part of the integrated site operation. The mobile system is moved on top of the already completed floors to the column positioned and aligned by the manipulator and then to the column to be fixed by welding (Figure 2.27). The robot embraces the column and welds it from two sides simultaneously, so that any distortion of the steel column by the welding process is avoided. Several welding robot systems can be used to weld more than one column in parallel.
7. *Welding robot system for beams.* The welding robot system for beams is positioned by the manipulator on top of preliminary fixed beams to weld/fix the beam components to the column components (Figure 2.26). The welding system consists of two frames used for the attachment to the beams and four welding robots, with one welding head each, mounted on top of these frames.
8. *Mobile (horizontal/floor-by-floor operating) logistics robots.* On floors within the SF that are already covered with ceiling elements, and that are thus no longer accessible by the OMs, mobile logistics robots (STCRs; **Volume 3**) are used for horizontal material delivery (mainly for delivering material for interior finishing). The logistics robots are integrated into the management and real-time control system and can communicate with the automatic vertical lifts to achieve JIT, JIS material delivery.
9. *Alignment and accuracy measurement system.* Sets of optical sensor systems (stereo vision systems, 1D and 2D laser scanning systems) are used to support the alignment and accuracy measurement.
10. *Real-time monitoring and management system.* Data from sensor systems as well as from the servomotors/encoders of the V/HDS along with information obtained from cameras monitoring all activities (including human activities) in the factory are used to create a real-time representation of equipment activity and construction progress (Figures 2.16 and 2.34). A barcode system (Figure 2.33) allows representation and optimization of the material flow and allows equipment (such as the automatic lift) to identify the component being processed. Real-time monitoring and management is done in a fully computerized on-site control centre.

End-effectors The manipulators (HDS) can be equipped with various end-effectors suiting the pickup, positioning, and adjustment of individual component types:

1. End-effector for pickup, positioning, and adjustment of columns (Figure 2.25)
2. End-effector for the horizontal delivery of multiple beams or steel profiles

3. End-effector for the positioning of beam/girder elements (Figure 2.28)
4. End-effector for pickup and positioning of concrete floor slabs (Figures 2.29 and 2.30)
5. End-effector for delivery and positioning of a welding robot system for beams
6. End-effector for positioning of prefabricated large-scale facade elements (Figures 2.31 and 2.32)
7. Jigs and fixtures for material delivery (Figures 2.42, 2.43, 2.44, 2.45, 2.46, 2.49, 2.50, and 2.51)

System Variations The SF consists of a modular steel frame system that can be adjusted to buildings with varying lengths and widths. Depending on the width of the building, one to three rail systems (and thus manipulators/HDSs) can be installed side by side. With the rising amount of HDSs, multiple vertical lifts can be installed inside the floor area or along the facade. So far only rectangular buildings have been manufactured with ABCS. The end-effectors have an inbuilt flexibility that allows them to pick up components of the same type with varying diameters or dimensions. The possibility of exchanging end-effectors allows completely new and initially unforeseen types of components to be used later. During the application of ABCS in various projects, Obayashi experimented with different site-structuring and automation ratios:

1. ABCS: High ratio of systematization and automation, used for erection of buildings with steel frames
2. Hybrid-ABCS: Combination of conventional construction with ABS, used for erection of buildings with steel frames
3. Big Canopy: Variation and recombination of ABCS technologies, used for erection of concrete buildings

Robot-Oriented Design The system's vertical and horizontal manipulators process mainly prefabricated low-level components (steel columns, steel beams, concrete floor slabs) and high-level components (facade elements completely equipped, e.g., with windows). All components are preconfigured to fit into the manipulators' working space and DOFs, and to the end-effector range/inbuilt flexibility. The alignment of steel columns, steel beams, and concrete floor slabs is guided by compliant joining systems, and the installation of facade components by compliant and plug-and-play joints, reducing the need for human workers to guide or assist in the installation process.

Efficiency Analysis

ABCS has a high ratio of systematization and automation, used for erection of buildings with steel frames. Obayashi captured a tremendous amount of data during the operation of the system, and, on this basis, analysed the impact of productivity, efficiency, working conditions, and building quality.

Rough Project Schedule ABCS, and its variation the Hybrid-ABCS, have been used a total of five times commercially (Ikeda & Harada, 2006; Figure 2.2):

- Project 1 (ABCS): 10 floors, 650 m^2 per floor; 8 floors were built by the system
- Project 2 (ABCS): 26 floors, 2700 m^2 per floor; 19 floors were built by the system

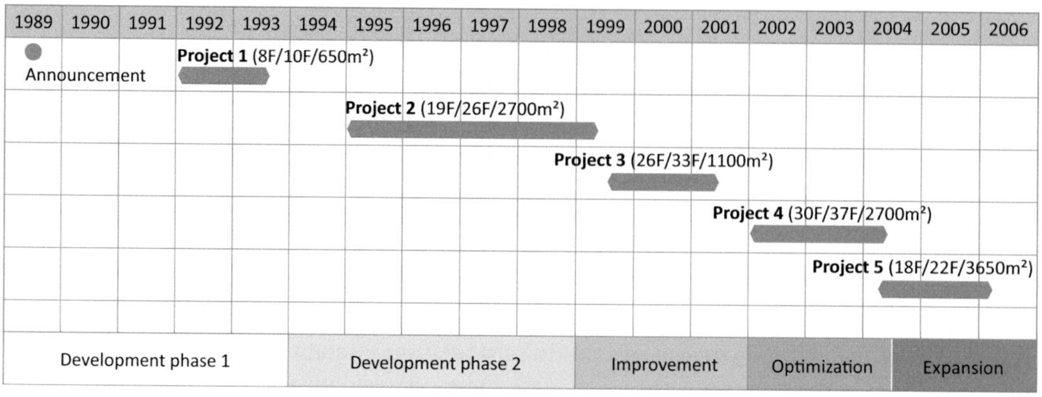

Figure 2.2. Comparison of time needed by different ABCS applications in different projects for the production of the building (adopted from Ikeda & Harada, 2006).

- Project 3 (ABCS): 33 floors, 1100 m² per floor; 26 floors were built by the system
- Project 4 (ABCS): 37 floors, 2700 m² per floor; 30 floors were built by the system
- Project 5 (Hybrid-ABCS): 22 floors, 3650 m² per floor; 18 floors were built by the system

Projects 3 and 4 showed, in particular, that the actual construction time, including interior finishing by ABCS, could be reduced to nearly 6 months, which, considering the amount of floors erected in these projects, is an extremely short amount of time. The reduction of construction time during these projects is even more remarkable considering the fact that the system was still in a state of optimization and improvement, and that during each application Obayashi experimented with new configurations of the system, so that learning effects across projects could not have been deployed fully. It turned out that general planning, and execution planning in particular (which can be largely related to the planning of the application of ABCS), took more time than is usual in conventional construction, owing to the use of ABCS. In conventional construction, general planning, and execution planning in particular, are often both partly parallelized with the construction process. As ABCS cannot be operated fully in the lower two to four floors and top one to three floors, but only in those in between, the number of floors that can be constructed by ABCS is always lower than the total number of floors by a certain amount. Related to the total amount of floors in the building that ABCS allows is the fact that buildings are constructed at a speed of three floors (buildings with larger floor areas) to four floors (buildings with smaller floor areas) per week.

Floor Erection Cycle The floor erection cycle (FEC) of ABCS follows a 6-day schedule (Figure 2.3). Work is done on several levels in parallel. On the uppermost floor (*n*th floor) in the SF, the steel frame erection is done by the automated OMs for the middle column row. At the same time, the two automatic OMs operating from the outer column rows install exterior facade elements on the lower levels ($n - 2$ floor). After 3 days, the middle column row is completed and the SF can be lowered by

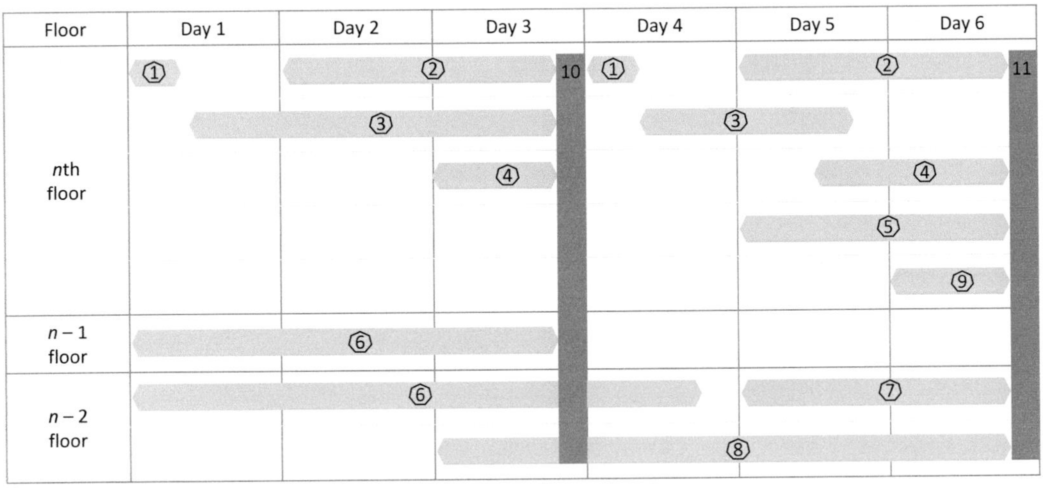

(1) Column positioning and adjusting; (2) Column welding; (3) Beam positioning, adjusting, and fixation (bolting); (4) Precast concrete element installation; (5) Floor finishing; (6) Exterior precast concrete panel installation; (7) Glazing; (8) Insulation and caulking; (9) Climbing of the VDS; (10) SF/CS lowering (700mm); (11) SF/CS climbing

Figure 2.3. Floor erection cycle for standard floors and with system in full operation (adopted from Miyakawa et al., 2000).

70 cm to rest statically on those columns. The outer column rows can then be built. As the SF enables work on the facade (installed on $n-1$) on $n-2$ to take place from the inside and outside of the building, on a safe and enclosed working platform, the facade on level $n-2$ is sealed and finished completely within the SF. Interior finishing is then done on the lower floors of the (sealed) building.

Experiments with Degree of Automation/System Configuration/Worker Teams
During the application of ABCS in various projects, Obayashi experimented with different automation ratios. The experiments, following the optimum principle, tried to find an optimal relation between a certain automation ratio and the output performance. The experiments focused on the automation of the OMs and the welding system (Miyakawa et al., 2004):

- *Operation of OMs in SF.* In some projects, the OMs were operated automatically only at a location close to the installation location and final positioning was done manually by remote control. In other projects, the whole logistics and positioning process of VDS and HDS was automated. It turned out that the first approach was more cost effective, as faster operation was possible. Full automation turned out to be slower owing to real-time processing of localization, motion planning, fusion of sensor data, etc.). Furthermore, the first approach needed less effort for programming and was less error prone.
- *Welding operation.* Within the ABCS applications, Obayashi conducted experiments with welding robots. Welding robots turned out to be slow and in need of continuous supervision. In addition, they could be used only for simple welding operations (e.g., straight line around column). Human welders were faster in conducting simple welding operations. For more complex welding operations they were, however, a necessity.

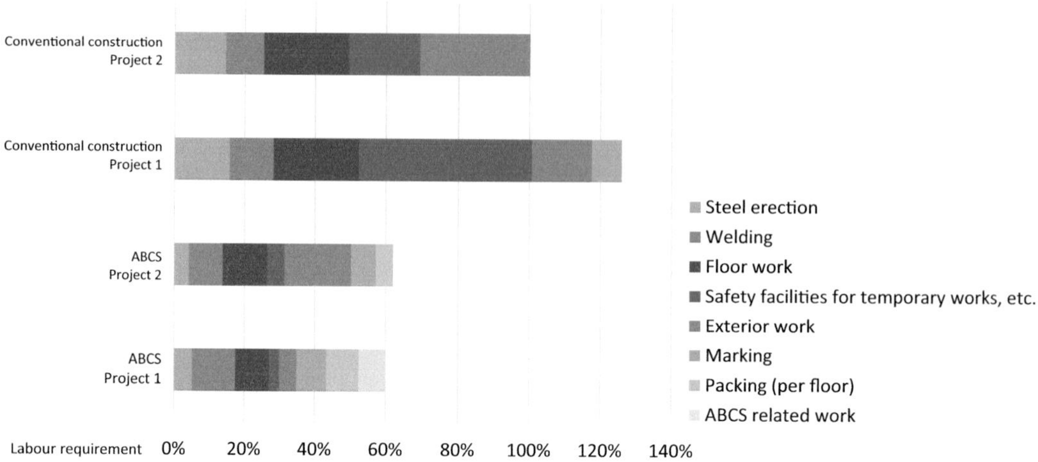

Figure 2.4. Comparison of labour requirements per floor (adopted from Miyakawa et al., 2000).

Productivity Compared to conventional construction, ABCS is able to reduce the average labour requirement (man-hours) per floor by around 40%. Reductions are possible in nearly all work categories. High reductions of man-hours have been achieved regarding the positioning of steel frame elements (steel columns and beams), floor installation work, and the installation of safety facilities (Figure 2.4). The application of ABCS has also created new work categories on-site.

Learning Effects Obayashi has found that within task categories the time necessary to complete a task in average during the project decreased by 10% (Miyakawa et al., 2004).

Product Quality and Process Monitoring To be able to survey and record all construction operations, and thus to guarantee the quality of the final product, ABCS uses an on-site control centre with the following capabilities:

- Surveys equipment operation in real time (cameras, sensor data, motor data) – visual real-time representation of equipment operation
- Barcode system manages material during delivery, logistics, and positioning process
- Determines order of assembly
- Controls work schedule and supply of material
- The system reduced the need for on-site construction management and conventional quality control methods

Safety ABCS lowers the physical strain and improves the health and safety of the workforce (Miyakawa et al., 2004):

- It provided safe working platform for workers. As ABCS always creates a full floor, including floor slabs (in steel building erection, the frame was often first completely erected). The lower floor always serves as a safe working platform during steel column and beam assembly (e.g., welding operations).

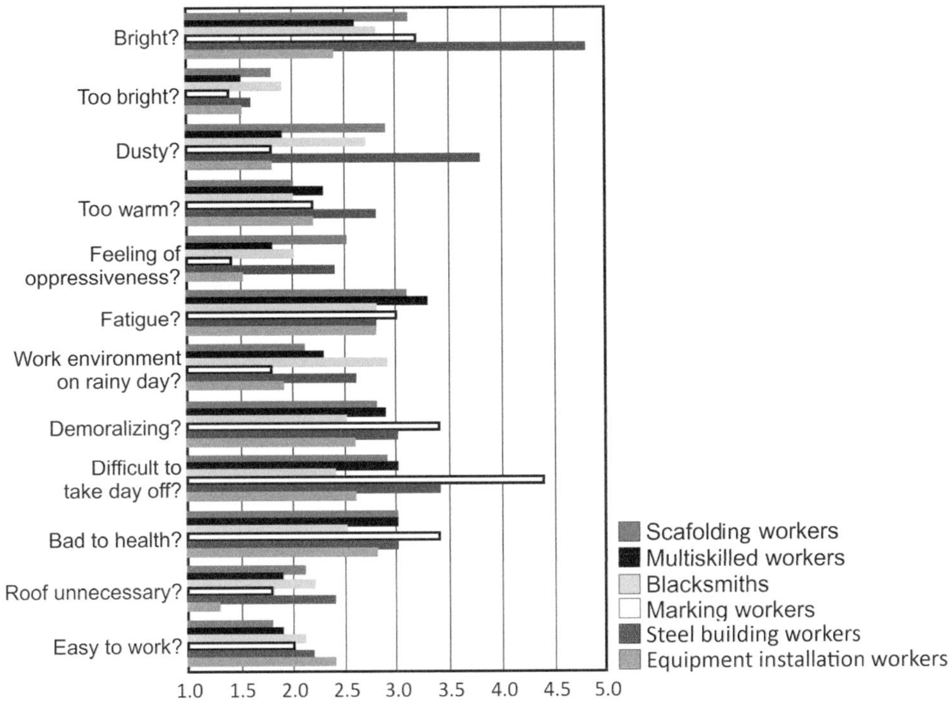

Figure 2.5. Rating of usability of features of ABCS (in particular SF) approach. Comparison of rating outcome for each feature according to various classes of workers; a rating of 5 indicates a negative opinion, whereas a rating of 1 indicates a positive opinion (adopted from Miyakawa et al., 2000).

- Safe working platforms for outside facade work are provided by the SF (no scaffolding necessary).
- The OM operators sit in the SF with a view over the whole area, and are thus able to work close to the installation location. This enhances accuracy and safety, improves teamwork, and reduces the possibility of injuring someone by the OM or during material lifting.

Usability During application of ABCS, Obayashi conducted usability studies and analysed how the system influenced the workers' behaviour, emotions, and motivations related to the system (Miyakawa et al., 2000; Figure 2.5). The study also compared different categories of workers. In general, the workers evaluated ABCS more positively than negatively (most ratings were above average). However, the study also revealed that the systemized ABCS environment made workers think that it is more difficult to take a day off (thus many reported feeling fatigued). Furthermore, the lighting and the bright working environment in the ABCS SF was not evaluated positively. All in all, it can be concluded that the systemized ABCS, from a worker's point of view, obviously reduces the ability for own decisions regarding breaks and individual work speed; it changes the workflow and thus habits of the construction workers; and as the usability studies have shown, this is not viewed positively in all aspects. This phenomenon accompanies many new developments that involve technologies that are able to track or record people and processes.

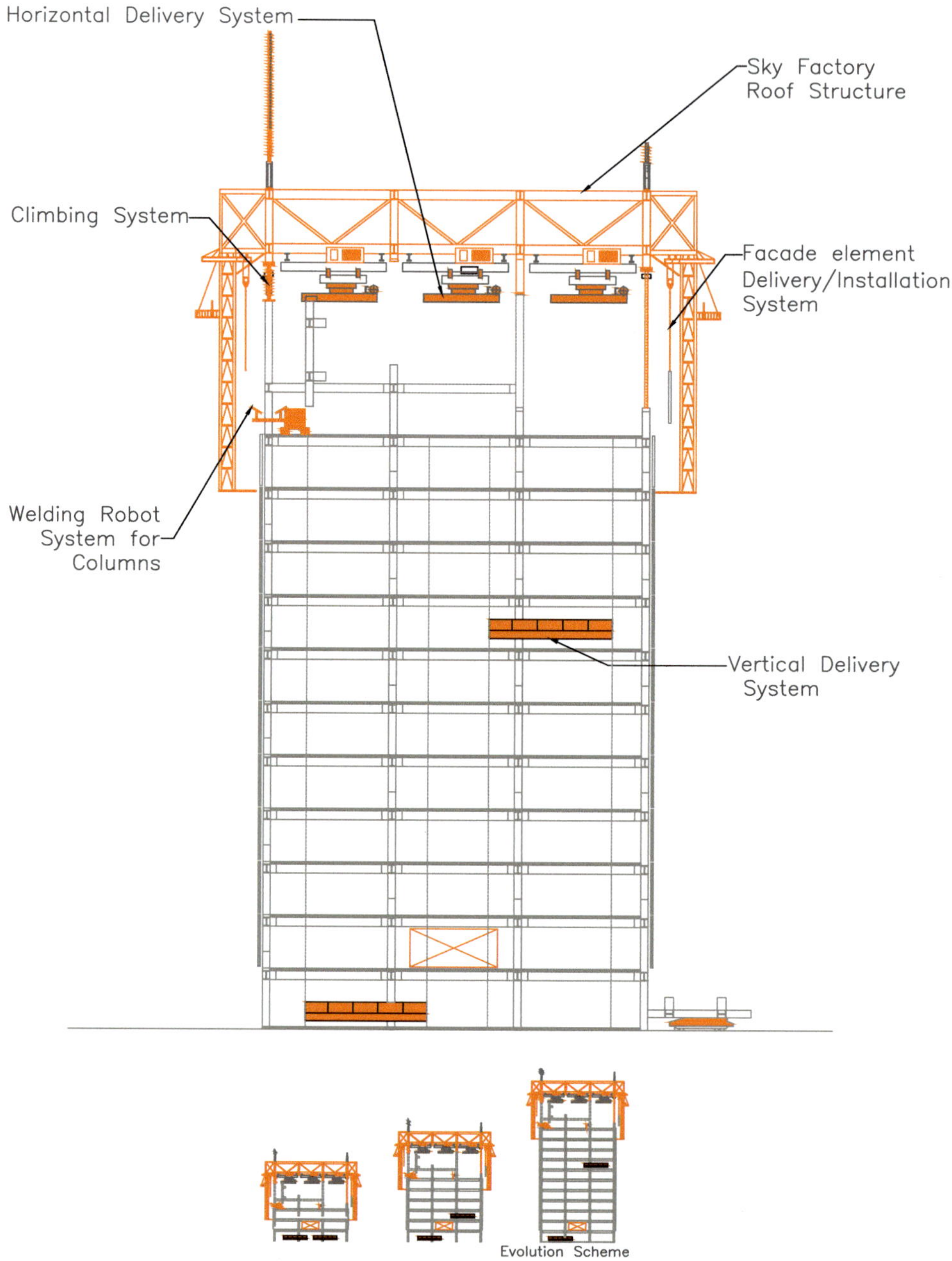

Automated Building Construction System (ABCS), Company: Obayashi Corporation, Japan

ABCS Application Example 1

Figure 2.6. Sky factory front view.

Figure 2.7. Vertical delivery system delivering material to a sky factory.

Figure 2.8. Sky factory side view.

Figure 2.9. Sky factory interior view; overhead manipulator (used for column installation) travelling along rails.

Figure 2.10. Overhead manipulator and climbing system.

Figure 2.11. Climbing system, stilt.

Figure 2.12. Overhead manipulator.

Figure 2.13. Overhead manipulator.

Figure 2.14. Overhead manipulator.

Figure 2.15. Overhead manipulator.

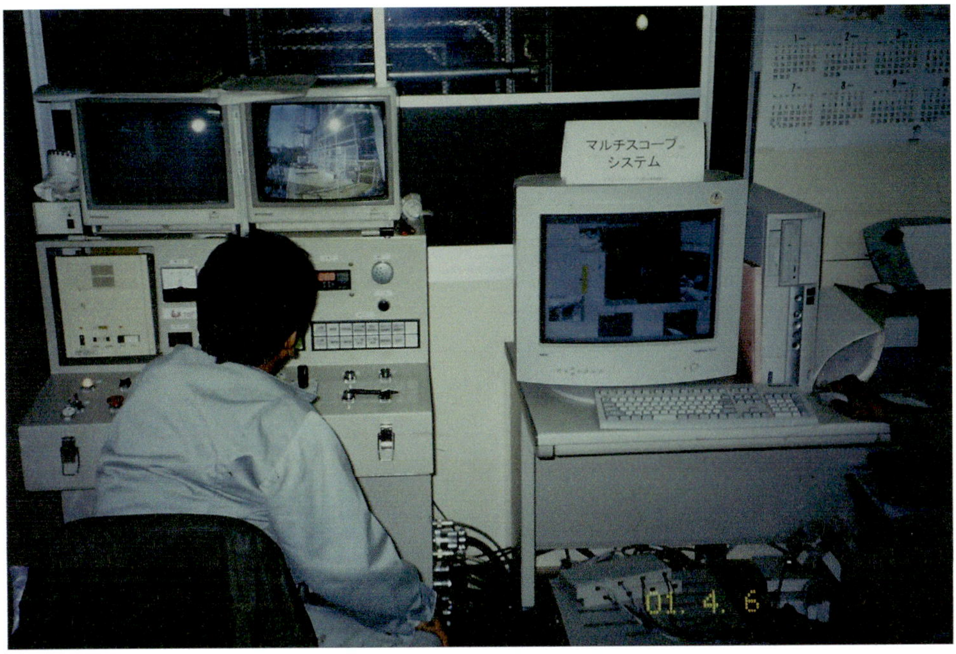

Figure 2.16. On-site control room.

Figure 2.17. Building component (here steel beam) is designed for optimized manipulation and assembly.

ABCS Application Example 2

Figure 2.18. Sky factory exterior view.

Figure 2.19. Sky factory exterior view.

Figure 2.20. Sky factory exterior view.

Figure 2.21. Finished building.

Figure 2.22. Sky factory exterior view.

ABCS Application Example 3

Figure 2.23. Pick-up of column by a vertical delivery system on the ground. (Image: Obayashi, documentary video)

Figure 2.24. Pick-up of column by an overhead manipulator on operational floor. (Image: Obayashi, documentary video)

Figure 2.25. Positioning of column by an overhead manipulator. (Image: Obayashi, documentary video)

Figure 2.26. Beam welding system. (Image: Obayashi, documentary video)

Figure 2.27. Column welding system. (Image: Obayashi, documentary video)

Figure 2.28. Positioning of beam by an overhead manipulator. (Image: Obayashi, documentary video)

Figure 2.29. Pick-up of floor slabs by an end-effector. (Image: Obayashi, documentary video)

Figure 2.30. Positioning of floor slabs by an end-effector. (Image: Obayashi, documentary video)

Figure 2.31. Positioning of façade element. (Image: Obayashi, documentary video)

Figure 2.32. Compliant plug-and-play connector façade element. (Image: Obayashi, documentary video)

Figure 2.33. RFID-based supply and on-site logistics management. (Image: Obayashi, documentary video)

Figure 2.34. Real-time monitoring of manipulators and construction progress. (Image: Obayashi, documentary video)

ABCS Application Example 4

Figure 2.35. Material delivered into sky factory through a material intake port by an overhead manipulator.

Figure 2.36. Side view of a sky factory, early construction stage.

Figure 2.37. Side view of a sky factory, late construction stage.

Figure 2.38. Climbing system, lower end of climbing stilt.

Figure 2.39. Climbing system, raised and lowered climbing stilts.

Figure 2.40. Climbing system, exterior view.

Figure 2.41. Sky factory, detailed view of material intake port.

Figure 2.42. Jig used to present concrete components to overhead manipulators and their end-effectors.

Figure 2.43. Jig used to present concrete components to overhead manipulators and their end-effectors.

Figure 2.44. Sequencing of construction material on the construction site.

Figure 2.45. Kanban principles used on the construction site.

Figure 2.46. Automated on-ground material logistics.

Figure 2.47. Interior view of a sky factory.

Figure 2.48. Sequenced column components; view of the compliant joining system.

Figure 2.49. Ground factory.

Figure 2.50. Construction material picked up in ground factory.

Figure 2.51. Logistics jig used to transport column components.

2.1.2 Akatuki 21

Company: Fujita Corporation, Japan
Category: SF (moving upwards) supported by building

Analysis of Technology and Composition

Evolution Scheme Akatuki 21 distributes work activities over two separated (but closely linked) factory parts: GF and SF. In the GF, material is stored, processed (e.g., parts are assembled into components), and prepared for final assembly in the SF. The GF reduces amount and complexity (amount of assembly steps, necessity of complex kinematics) of operations necessary in the SF. The SF is the focal point of the final assembly (Figures 2.52–2.54). It provides an enclosed and structured environment (SE) for the installation of components by rail-guided gantry type OMs. The SF operates on a FEC of 6–7 days. On the main operational level (*n*th floor), the

Figure 2.52. Exterior view on the sky factory of the Akatuki 21.

installation of column/beam components, and concrete floor slabs, and distribution of and levelling of concrete is done based on the OMs. On the floors below this level ($n - X$ floors), interior finishing is done in parallel.

Elevation The analysis reveals the concentration and interlinking of automated devices within GF and SF. Various VDSs connect the two factory parts, building up a continuous material flow between both. One large automatic lift (located centrally within the manufacturing area) supplies all parts and components (VDS1). A separate lift directly supplies and orientates beam components prepared in the GF towards an OM equipped with an end-effector for beam component handling and installation (VDS2). During delivery, each beam component is brought by the lift from a horizontal position into a vertical position. A further VDS, (VDS3) located at the outer roof of the SF, is used to lift material directly into the SF. A jib crane on top of the SF supports vertical logistics and is used to assemble, maintain, and disassemble the SF and other subsystems.

Ground Plan In the GF, both rail-guided automated guided vehicles (AGVs) and gantry-type OMs are used to transport components from the stock yard or the material handling, sorting, and processing yard (MHSPY) to the various VDSs. In the GF materials are unloaded from transport trucks, sequenced, and are quickly transferred, for example, via rail-guided AGVs (if necessary JIT, JIS) to a VDS or the MHSPY. In the SF, depending on the configuration of the building, one or two rail systems are installed so that along this rail systems, the OMs can move, pick up, and handle components supplied by the VDSs, and position and adjust them.

Subsystems Akatuki 21 manufactures the building on-site by GF and SF and consists of a set of integrated subsystems:

1. *Ground factory delivery yard.* Trucks deliver incoming materials to an enclosed SE erected on a ground level and attached from the outside to the building. Overhead cranes assist in unloading material to AGVs for precast concrete panels conveyance. Furthermore, various other devices for the handling of parts are used (Figure 2.59).
2. *Ground factory parts/components stock yard.* Enclosed factory part erected on ground level and attached from the outside to the building. Used for warehousing.
3. *Ground factory parts/components MHSPY.* Enclosed SE erected on the ground level and attached from the outside to the building used for preparing components for the delivery and SF installation process. Parts and lower level components are transformed into higher level components.
4. *Ground factory horizontal delivery system.* AGVs transport material between the different GF yards and from the yards to the different VDSs.
5. *Vertical delivery system 1 (automatic lift).* An automatic lift operates in a shaft within the manufacturing area to transport all types of parts and components to the *n*th floor where the OMs operate.

6. *Vertical delivery system 2 (column lift)*. An automatic lift supplies columns vertically into the GF (parts/components processing yard) in a horizontal position and delivers them guided by rails on a shuttle to the SF in a vertical position (Figure 2.64). The horizontal position of the components on the ground level simplifies delivery and preparation; their vertical position in the SF simplifies the pick-up and installation process.
7. *Vertical delivery system 3*. Vertical exterior lift. On their delivery, the lift assists workers with the handling and installation of exterior components such as facades elements.
8. *Sky factory structure*. The SF uses the ceiling of one of the top floors as a roof that encloses the work area and to which the rail systems guiding the manipulators are mounted (Figures 2.57, 2.63, and 2.72). Up to two top floors are conventionally constructed on the ground level and are pushed up as the SF rises, finally becoming part of the building.
9. *Climbing system*. The SF (or rather the two top floors) are supported and lifted hydraulically by stilts that rest directly on the erected columns. The stilts are lifted one by one to place new columns under it. Once all columns have been placed, the factory is pushed up by hydraulic actuators.
10. *Sky factory horizontal delivery system*. The gantry-type OMs with exchangeable end-effectors in this system are rail guided (Figures 2.55 and 2.68). With three DOFs the OMs are able to reach any point within the SF. In combination with the end-effectors, additional DOFs are achieved. Depending on the width of the building one or two rail systems (and thus OMs) can be installed side by side. The horizontal SF delivery system is directly integrated with the VDSs so that an uninterrupted material flow is possible.
11. *Welding robot system for columns*. The welding robot systems for columns are mobile STCRs used in the SF (Figures 2.58, 2.69, and 2.70). The mobile robot system is moved on top of the already completed floors to the column (positioned and aligned by the OM) to be fixed by welding. The robot embraces the column and welds it simultaneously from two sides so that any distortion of the steel column by the welding process is avoided. Several welding robot systems can be used to weld columns in parallel.
12. *Real-time monitoring and management system*. Fujita uses a control room to supervise work tasks, material flow, progress of construction, quality, and the operation of all the subsystems (Figure 2.71).

End-effectors The manipulators (HDS) can be equipped with various end-effectors to suit the pick-up, positioning, and adjusting of various component types:

1. End-effector for pick-up, positioning, and adjustment of columns (Figure 2.65)
2. End-effector for positioning of beam/girder elements (Figure 2.66)
3. End-effector for concrete placing and concrete levelling (Figures 2.56 and 2.67)
4. End-effector for positioning of prefabricated large-scale facade elements (Figures 2.60–2.62)

System Variations Within the rectangular shape and orthogonal organization of the SF, buildings can vary in width and length. The grid structures of the building, GF, and SF have to be in tune with each other. The detachability of the end-effector systems allows new building component types to be added and efficiently handled by co-adjusted end-effectors without the need for major changes to the OMs or other subsystems.

Robot-Oriented Design Akatuki 21 follows the principle of robot-oriented design (ROD) on a component level, that is, by the design of the column components that allow for fast assembly as well as for an efficient handling by VDS and OMs. On a building level, the efficiency can, for example, be enhanced by designing buildings that are longer than they are broad to better exploit the capacity of the rail-guided part of the OMs. On a city level, the optimal functioning of logistics and the GF (interplay of storage yard and MHSPY elements) can be supported by an appropriate allocation and orientation of the building.

Efficiency Analysis

Akatuki 21 was developed by Fujita for the construction/assembly of steel-based buildings on the construction site. Unlike many other systems, Akatuki 21 consists of a GF (material delivery, preparation, warehousing and logistics, joining of lower level parts/components to elements) and an SF (with automatic/robotic OMs for element installation). Through the deployment of a GF that assembles parts into components, Akatuki 21 is more efficient than other systems in terms of the logistics from the factory/suppliers to the site, as parts (which are processed on site in the GF into components) in general can be shipped to the site more efficiently than large components.

Rough Project Schedule To set up Akatuki 21, Fujita first installs the SF and the GF on-site, which takes about 1–2 months. The installation process takes longer compared to other systems, as the SF also represents the two uppermost floors of the building. These two floors are completely assembled on the ground (including facade finishing) and serve as the "roof" of the SF to which the equipment (OMs including hoist/end-effectors and guiding rail system) is fixed. The SF then starts operating from the second floor onwards and assembles all of the following floors. In full operation, Akatuki 21 can construct around three floors per month, so that about 4 months are needed to construct, for example, 11 floors for a building with 15 floors. The disassembly of the equipment of the SF and the GF after completion of the building requires about 1 month.

Floor Erection Cycle Akatuki 21 operates with an FEC of 8 days. Day 1 is used to install the column components. Over days 2–3, the beam components are installed. Further beam parts are installed on day 4, and the concrete floor components are positioned on days 5–6. Concrete is poured and levelled on day 7, and finally on day 8 the SF is lifted and prepared for the construction of the next floor. For each different operation the OMs are equipped with the necessary end-effectors.

Product Quality and Process Monitoring Fujita uses a control room to monitor quality relevant aspects and guarantee the correct assembly of the product.

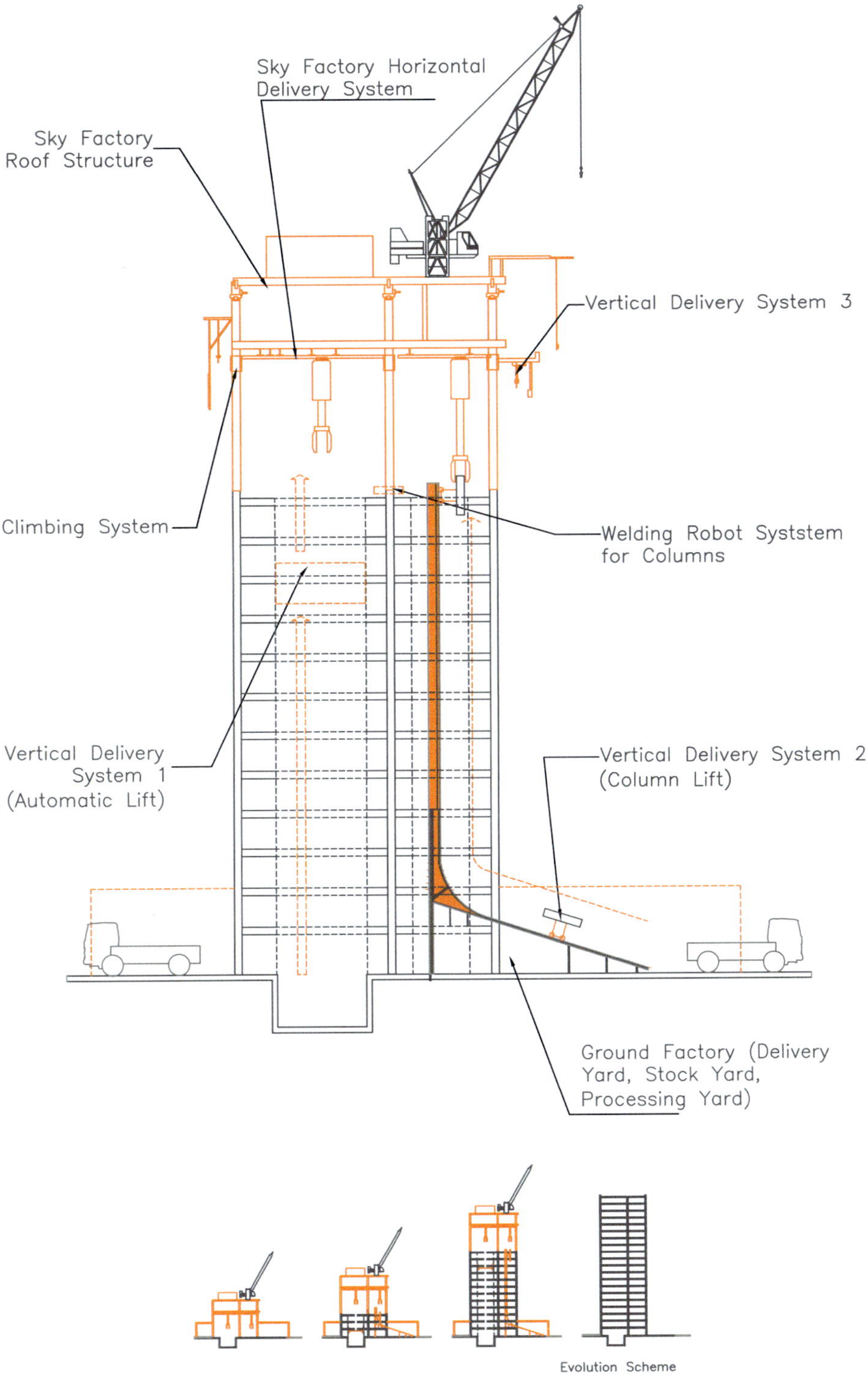

Akatuki 21, Company: Fujita Corporation, Japan

Figure 2.53. Exterior view, early construction stage.

Figure 2.54. Finished product.

Figure 2.55. Sky factory, interior view; overhead manipulators equipped with end-effectors for column and beam assembly.

Figure 2.56. Sky factory, interior view; overhead manipulators equipped with end-effectors for concrete distribution and levelling.

Figure 2.57. Sky factory, interior view; view of rail system guiding the overhead manipulators.

Figure 2.58. Robot system used for welding. (Image: Fujita Corporation)

Figure 2.59. On-ground logistics.

Figure 2.60. Temporary overhead manipulation systems are installed on floors under the sky factory for supporting interior finishing.

Figure 2.61. Optional robot system for installation of facade components. (Image: Fujita Corporation)

Figure 2.62. Jig used to guide faceded components into place. (Image: Fujita Corporation)

Figure 2.63. Outside view of a variation of the sky factory. (Image: Fujita Corporation)

Figure 2.64. Disassembly of the sky factory. (Image: Hasegawa, 1999)

Figure 2.65. Column components are handed over from the vertical delivery system to the overhead manipulator's end-effector for column component installation. (Image: Hasegawa, 1999)

Figure 2.66. Overhead manipulator equipped with end-effector for beam component installation. (Image: Hasegawa, 1999)

Figure 2.67. End-effector for concrete distribution and levelling. (Image: Hasegawa, 1999)

Figure 2.68. End-effector for column/beam installation.

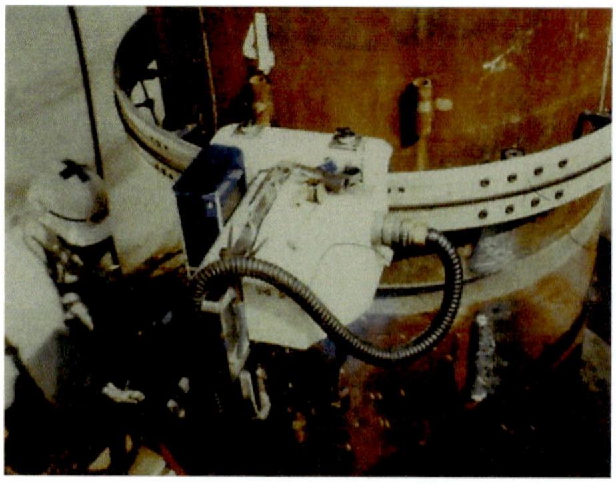

Figure 2.69. Automated column welding. (Image: Fujita Corporation)

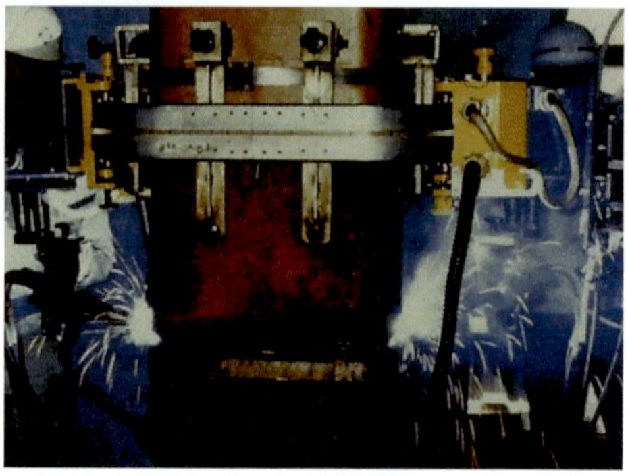

Figure 2.70. Automated column welding. (Image: Fujita Corporation)

Figure 2.71. Control room. (Image: Hasegawa, 1999)

Figure 2.72. Sky factory, detailed view.

2.1.3 Future Automated Construction Efficient System (FACES)

Company: Goyo/Penta Ocean, Japan
Category: SF (moving upwards) supported by building

Analysis of Technology and Composition

Evolution Scheme FACES concentrates work activity in an SF (Figure 2.73). The SF and the subsystems operating in it are assembled on the ground level. The SF, because of its configuration, starts operating on the third floor. Basement work and the erection of the first and second floors can be done in parallel with the SF set-up (Figure 2.88). The SF is composed of a roof, support frames, and sound mitigating

Figure 2.73. Exterior view of a sky factory.

side plates. In the uppermost floor, under the SF, OMs assemble the main structure (columns, beams, floor components). In the three floors under this operational floor (*n*th floor), fixation and welding is completed, facade elements are assembled, and interior finishing is done in parallel. Central to the system are the two OMs (Figure 2.80) accomplishing the vertical/horizontal component logistics and the positioning and adjustment of low- and high-level components, such as steel columns, beam structures, and facade elements.

Elevation The FACES SF completely covers the work area and provides an SE for main structure erection, facade work, component assembly and welding, and interior work. In total, four floors are considered to be part of the SF: the *n*th (uppermost) floor, $n - 1$ floor (welding of girders and columns, concrete slab placement), and $n - 2$ and $n - 3$ floor (e.g., interior finishing, facade sealing and finishing). The lead times per floor and the FEC are analysed in detail in the *Efficiency Analysis* that follows. Once the construction of a floor is completed, the system is pushed up by means of a CS (powered by 250-ton oil-hydraulic actuators at a rate of 2 mm/s, that is, approximately one floor per hour). This process is repeated floor by floor until the whole structure is completed and the SF can then be disassembled (Figure 2.89).

Ground Plan The analysis of the ground plan reveals that the two high-capacity OMs operating along length direction of the factory play a central role. Each OM operates in a fixed trajectory along two rail profiles mounted to the supporting roof frame of the SF. To transfer components from the ground yard to the target position, the SF has two material intake ports: the main intake port and sub-intake port (Figure 2.79). To raise the SF, the CS pushes it up on steel stilts (in this case a steel truss embracing a column) resting on top of the columns installed by the manipulators beforehand. The CS is positioned only over certain columns (e.g., every second one) and functions also as a fixture/template that guides columns into their adjusted position and ensures the precise assembly of other columns by giving reference points.

Subsystems FACES, manufacturing the building on-site in an SF, consists of a set of integrated subsystems:

1. *Material handling, sorting and processing yard (optional).* Mobile cranes for material handling on the ground are used in the MHSPY (Figure 2.87).
2. *Vertical delivery system (optional).* Temporary lifts for delivery of parts/components to the SF.
3. *Sky factory structure.* The SF encloses the workspace on the *n*th to $n - 3$ floors completely, ensuring a weather-protected SE. It also serves as a carrying frame that integrates the guiding rails of the OMs and provides incoming ports for material and component delivery (Figures 2.77–2.79). The SF also provides working platforms for outside facade work. When the construction work is taking place on the *n*th floor, the wall installation task is carried out on the $n - 3$ floor. To position the exterior walls and other wall finishing, additional manual labour support is required, for which exterior platforms integrated into the SF are used.

Exterior sealing work can be finished from these platforms by workers once the wall components have been assembled by the OMs.

4. *Climbing System.* The CSs consist of steel truss-based stilts, which embraces the columns and support the SF (Figure 2.93). An inchworm-like pushing mechanism (Figure 2.94) operates in each steel truss. The lower end is fixed to the column and functions as a fixed point for the pushing up of the SF. The steel truss stilts of the climbing system also function as fixtures/templates for the positioning of column components by the OMs (Figure 2.90).
5. *Sky factory horizontal delivery system.* The HDS consist of gantry-type OMs with exchangeable end-effectors (Figures 2.80 and 2.81). The OMs are rail guided and the gantry system is able to move along two rails lengthwise, to rotate, to translate (extend main link), to rotate the last link, and to lower components along the axis of translation of the last link (5 DOFs). End-effectors are exchangeable and take over component-individual grasping/releasing functions. Each of the two OMs has a flexible and adjustable working radius (over the main link, which is extendable) of 6 m to 10 m. The two heavy but high-capacity OMs are key to the assembly process. Unlike in other systems, the two OMs have a high payload capacity, integrate a multitude of DOFs, and have a relatively large working space. Their stable (but heavy) composition and structure guarantees high precision during assembly operations. Furthermore, the OMs can also be utilized as a VDS, lifting up components from the ground (via the two material delivery ports) and thus eliminate the need for the handover of components from the VDS to HDS. A disadvantage is their low speed (in travelling along the overhead rails), also discussed in the *Efficiency Analysis* that follows.
6. *Welding robot system for columns.* Welding robots connect two steel column components vertically. The robots travel along guiding rails temporarily fixed to the columns (Figure 2.86). By precisely welding from opposite sides, they optimize the welding quality.
7. *Real-time monitoring and management system.* A site office (Figure 2.83) controls the SF and MHSPY through real-time monitoring of the construction progress, work activity supervision, analysis of the CS and possible distortions of the building's and the SF's structure, and real-time visualization of each manipulator's operations.

End-effectors The manipulators (HDS) can be equipped with various end-effectors to suit the picking up, positioning, and adjustment of various component types:

1. End-effector for picking up, positioning, and adjustment of columns (Figure 2.82)
2. End-effector for the transporting and positioning of multiple beam/girder elements (Figure 2.84). As the manipulators operate relative slowly in one operational sequence, up to three beams can be installed.
3. End-effector for the positioning of other nonstandardized parts and components
4. End-effector for curtain wall component installation (Figure 2.85)
5. End-effector for exterior wall component installation
6. End-effector for interior wall component installation

System Variations The system can be varied on the basis of a rectangular overall grid system in a lengthwise direction that synchronizes column and CS subgrids. The variation in the sideways direction is also possible considering that the manipulator's main link can translate by 4 metres, and that less overlapping of workspace means less manipulator capacity available in each workspace.

Robot-Oriented Design On a component level, the design of the components (size, dimensions, and load) is synchronized with the manipulator's workspace and kinematic capability, as well as with the end-effectors. A temporary alignment system (four small steel plates on each of the four column sides) is welded temporarily to each column to allow the columns to be guided into place (Figures 2.91 and 2.92). On a building level, it has to be considered that the OMs operate relatively slowly in lengthwise direction. This means that overly long buildings are disadvantageous, or might necessitate additional OMs. On the other hand, buildings that are not long enough will not take full advantage of the capacity of a single manipulator. Furthermore, the structure of the building should take into account the need to synchronize the column grid with the factory structure's grid system to be able to allocate the CS. On a city level, the design and orientation of the building and its surroundings should take into account the location of the logistics routes and that incoming material is delivered to the SF via dedicated ports.

Efficiency Analysis

FACES is used to erect steel-based buildings. The on-site factory that operates on top of the building is characterized by the two powerful OMs (also called "shuttle cranes" by Goyo). After several applications, the system shows both advantages and disadvantages. One advantage is that the two powerful and heavy OMs are able to manipulate and position heavy elements with high precision (with higher precision than the OM systems of SMART and ABCS, for example). A disadvantage is the low operating speed of FACES's OMs and the strong vibrations their operation causes. Furthermore, owing to the reliance on the two powerful OMs, the system is less flexible and adaptable to varying structures and shapes of building than, for example, the SMART that, as a result of its composition of small gantry OMs and travelling trolley hoists, allows a higher variety of configurations.

Detailed Project Schedule Figure 2.74 shows the time schedule for the erection of a building with 23 floors in the Nihonbashi area in Tokyo by FACES:

- *Set-up*. The set-up of FACES took, in total, around 5 months for the project outlined in Figure 2.74. The long period was due to the fact that this project was the first application of the SMART and included the time necessary to certify the operational safety. Later, the time needed to set up the system was reduced to 1–2 months. Nevertheless, in all projects, as FACES took over operation from the third floor; the basement work and the erection of the first and second floors could be done in parallel to FACES set-up.
- *Operation*. The system was able to erect 23 floors over 8 months, or the equivalent to two-and-a-half floors per month.

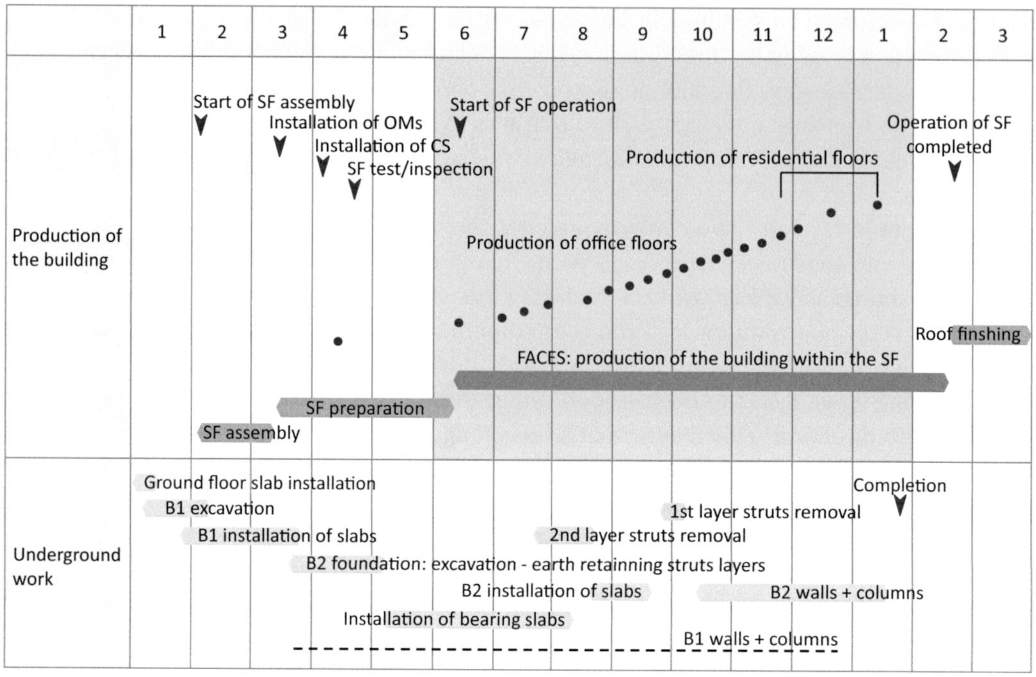

Figure 2.74. Overall work schedule for the erection of a building by FACES (adopted from Yoshida et al., 1997).

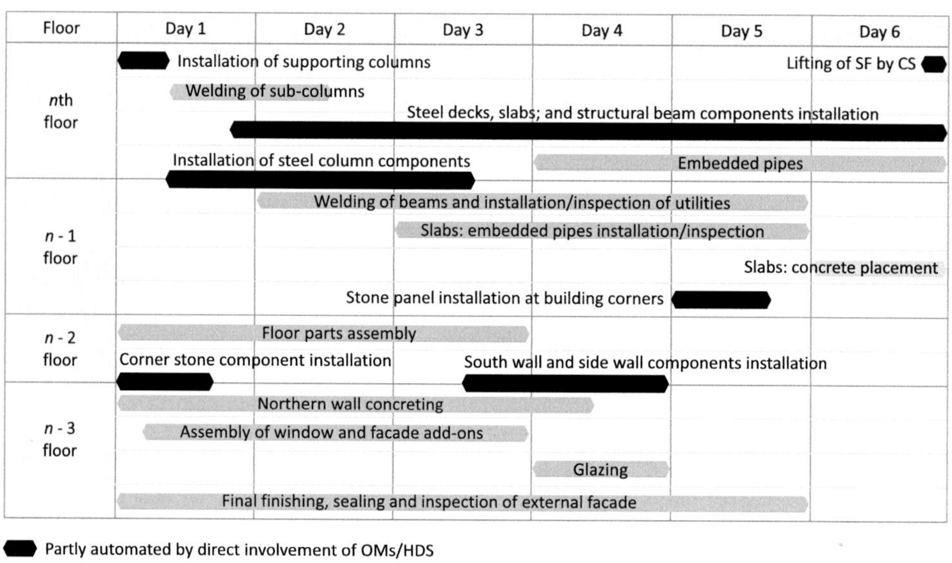

Figure 2.75. Floor erection cycle of FACES (adopted from Yoshida et al., 1997).

Table 2.1. *Overhead manipulator, specifications*

Rated suspension load: 10 tons	Travelling speed: 20 m/min
Boom turning radius: 6–10 m variable	Boom shuttle speed: 20 m/min
Lifting height: 120 m	Booming turning speed: 0.5 rpm
Lifting/lowering speed:	Hook turning angle/speed:
• 10 tons, 38 m/min	• Front: 90°/0.5 rpm
• 5 tons, 50 m/min	
• No load, 70m/min	
Two-hook independent operation speed/stroke: 0.45 m/min with a stroke of 0.5 m	

Adapted from Yoshida et al. (1997).

- *Disassembly*. The disassembly of the FACES took 1 month. During this time other finishing work took place.

During the phase where standard office floors had to be erected, FACES reached a maximum speed of three floors per month.

Floor Erection Cycle FACES worked in most projects with a floor erection cycle (FEC) of 6 days, meaning that the site factory was raised every 6 days (Figure 2.75). The FEC of FACES includes parallel processing on four floors (nth, $n-1$, $n-2$, $n-3$). The nth floor refers to the uppermost floor close to the OMs. Floors $n-1$ to $n-3$ are the floors below this uppermost floor. The FEC was designed around the OMs positioning operations.

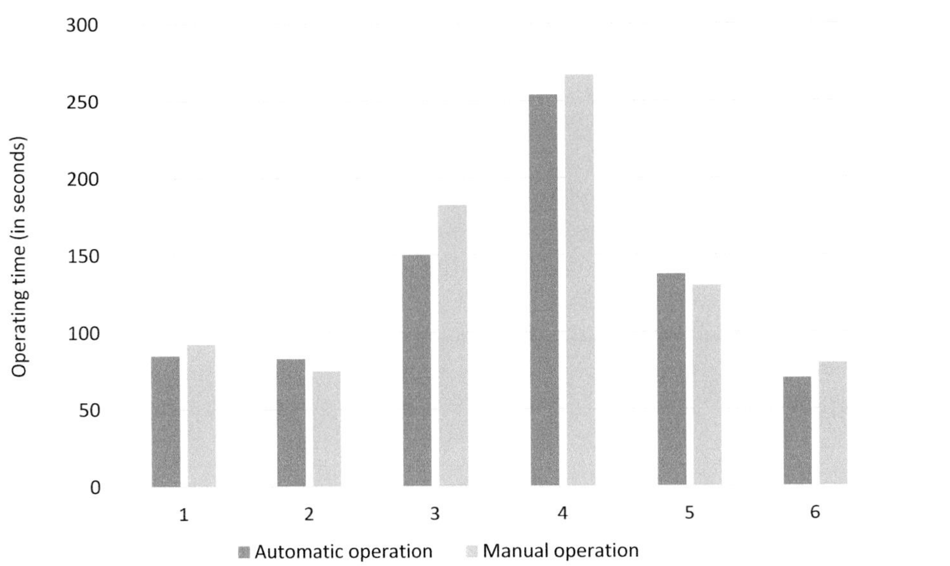

Figure 2.76. Comparison of automated/manual operation of the system related to the handling and installation of components (adopted from Yoshida et al., 1997).

Table 2.2. *Labor-saving result*

Working location (standard floor)	Efficiency of conventional methods	Efficiency FACES	Labour-saving factor at site
245 m²	**Knock-down system**	**Full unit system**	
Toilet room	2–3 m^2/man-day[a]	12 m^2/man-day	1/4–1/6
19 m^2, 23 m^2 rooms Moulding board installation	Conventional method 64 men/ standard floor	System toilet unit 15 men/standard floor	1/4
164 m²	**Nonscaffold method**	**Exterior work in SF (FACES)**	
Exterior sealing	3–4 m^2/man-day	6.4 m^2/man-day	2/3–1/2
Glass curtain wall	Using gondola 15 m^2/man-day	Exterior scaffolding work at the side shelter (FACES) 40 m^2/man-day[b]	1/3

[a] Frame assembly only, not inclusive of glazing and sealing work.
[b] Included instalment of water drainage pipes in 40 positions inside seals.
Adapted from Yoshida et al. (1997).

Technical Data, Speed of Equipment The OMs (Table 2.1) are the key element of FACES. Each OM has 5 DOFs, including prismatic and rotatory joints. Compared to other systems (e.g., Big Canopy or SMART with 30–40 m/min for gantry OMs and hoists) travelling and rotatory speeds are significantly lower (20 m/min). However, FACES's OMs are able to pick up and position several elements within one working procedure, thus reducing the need to travel long distances back and forth.

Experiments with Degree of Automation/System Configuration/Worker Teams Experiments concerning an optimized ratio of automation were conducted during the operation of FACES in the first projects. Different types of work were conducted in manual operation mode (remote controlled) and executed automatically by FACES's subsystems. The times needed to complete the work or tasks were recorded and compared (Figure 2.76).

Productivity FACES in the first projects was able to improve labour productivity considerably for work tasks that are usually work intensive (Table 2.2). Analysed work task fields showed an increase in productivity by two to six times. In addition, the improvement of work processes related to facade work in particular was able to improve the quality of construction and reduce the risk of injury.

Product Quality and Process Monitoring FACES uses a real-time monitoring and management system (RTMMS) that can help to set up, optimize, and supervise material flow, work tasks and processes. Manual control over the equipment can be executed via a computer. Automatic operation can also be supervised in real time. The information system links an off-site planning office with an on-site operations management office, a material handling yard, and the central control room at the FACES construction factory on top of the building. The system also assists with the acquisition and processing of data about completed work tasks on working and finishing floors (nth, $n-1$, $n-2$, and $n-3$).

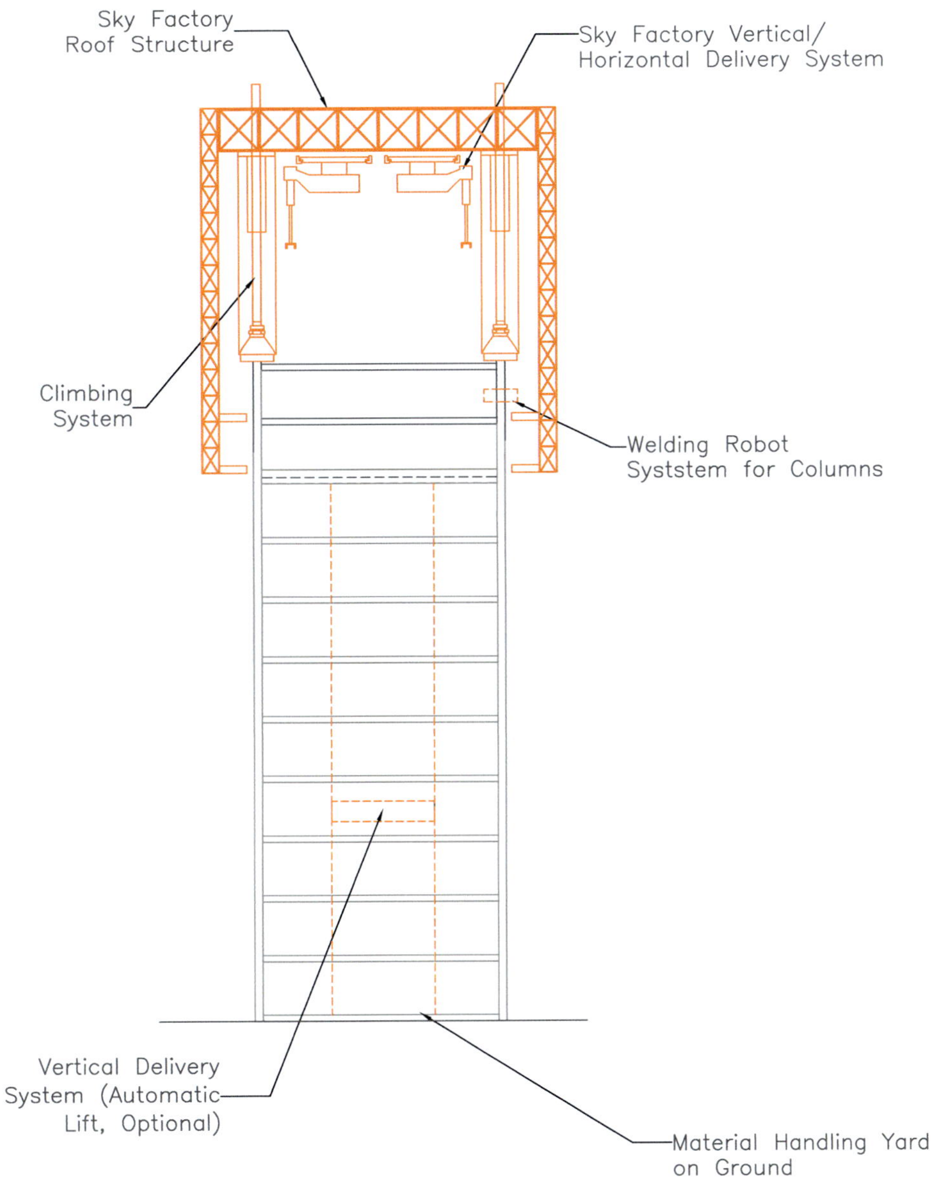

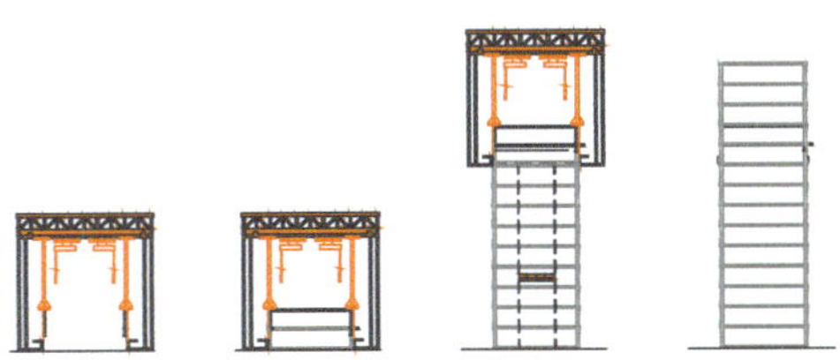

Evolution Scheme

Future Automated Construction Efficient System (FACES), Company: GoyoPenta Ocean Japan

Figure 2.77. Front view of sky factory.

Figure 2.78. Port for material intake.

Figure 2.79. Side view of a sky factory.

Figure 2.80. Interior view of a sky factory.

Figure 2.81. Detailed view of overhead manipulator. (Image: Goyo/Penta Ocean)

Figure 2.82. End-effector carrying column component. (Image: Goyo/Penta Ocean)

Figure 2.83. Control room.

Figure 2.84. End-effector carrying beam elements. (Image: Goyo/Penta Ocean)

Figure 2.85. End-effector carrying facade element. (Image: Goyo/Penta Ocean)

Figure 2.86. Automated column welding. (Image: Goyo/Penta Ocean)

Figure 2.87. Logistics gate of a ground factory.

Figure 2.88. Sky factory side view being built up on ground. (Image: Goyo/Penta Ocean, documentary video)

Figure 2.89. Sky factory disassembly. (Image: Goyo/Penta Ocean, documentary video)

Figure 2.90. Climbing system guiding column in place. (Image: Goyo/Penta Ocean, documentary video)

Figure 2.91. Column positioning. (Image: Goyo/Penta Ocean, documentary video)

Figure 2.92. Automated column alignment. (Image: Goyo/Penta Ocean, documentary video)

Figure 2.93. End-effector of climbing system. (Image: Goyo/Penta Ocean, documentary video)

Figure 2.94. Climbing system; detailed view of climbing mechanism. (Image: Goyo/Penta Ocean, documentary video)

2.1.4 Mast Climbing Construction System (MCCS)

Company: Maeda Corporation, Japan
Category: SF (moving upwards) supported by building

Analysis of Technology and Composition

Evolution Scheme The MCCS uses the steel structure of the uppermost floor of the building that it produces as the roof structure for the SF (Figure 2.95), to which other subsystems (e.g., OMs) are mounted. The roof structure is, therefore, built on the ground and then lifted (according to the progress of construction) floor by floor by means of steel stilts belonging to the CS. The assembly of the first few floors is done in the SE of the SF but without using the OMs (operational height not achieved for these floors). Full operation of the MCCS starts from the first or second floor. After the building assembly process is completed, the SF, CS and OMs are dismantled by two jib cranes positioned on the top of the building.

Elevation In the SF, two rail-guided OMs with partly overlapping workspace position and align the components of the load-bearing main structure of the building (consisting of low to medium level column and beam components) as well as large-scale facade components. The supply of material is accomplished via a VDS, which delivers material at the overlapping area between the workspaces of the two OMs.

Figure 2.95. Sky factory, exterior view.

In addition, each OM has access to a intake port on the side of the SF through which it can pick up and deliver components from ground directly. Assembly work takes place over three floors simultaneously (uppermost floor/*n*th floor: installation of column and beam elements; $n - 1$: installation of facade elements, plus interior finishing; $n - 2$: finishing and sealing of the facade from the outside by working platforms provided by the SF).

Ground Plan The ground plan outlines the grid system on which the columns are installed by the OMs. The SF rests on the movable steel stilts of the CS that sit on top of the outer row of columns on the two face sides of the building. For the positioning of columns under the stilts of the CS, the stilts have to be lifted. During the positioning process only the stilt under which a column has to be positioned is lifted and the others stay fixed to the building (fixed via column catchers) and thus carry the weight of the SF.

Subsystems The MCCS consists of a set of integrated subsystems:

1. *Vertical delivery system.* An automated lift operates in a shaft within the SF and the building and it is allocated in a way that allows both OMs in the area of their overlapping workspace to access it. The VDS is predominately used for the delivery of finishing material to the SF. A gate on the ground floor level allows one to identify components by barcodes and provides the OMs and workers with information about the installation process.
2. *Sky factory structure.* After completion of the automated assembly work the roof of the SF is used as the uppermost floor of the building. The SF's roof is built on the ground and then lifted up; the manipulators, side covers, working platforms, and so forth are then installed to it. The SF completely covers the building and provides a weatherproof SE (Figures 2.95–2.98). A control room is installed in the SF's roof.
3. *Climbing system and column catchers.* The SF roof structure is supported by steel stilts that rest on the (already installed) outer columns of the building (Figure 2.99). For the lifting process, automated hydraulic lifting mechanisms are connected to each stilt in the roof structure of the SF (Figures 2.105–2.107). At the bottom of each stilt an automatic "column catcher" is installed. This is a hydraulic device for temporary combining and fixing the stilt and the column of the building for supporting the SF.
4. *Sky factory horizontal delivery system.* The OMs are the central subsystem of the MCCS (Figures 2.100–2.104). Each of the two OMs has 5 DOFs: (1) translation of main trolley lengthwise along main rails attached to the SF's roof, (2) translation of sub-trolley (right/left) within the rails of the main trolley, (3) turning of main link, (4) extension of main link along the axis, and (5) lifting/lowering of end link. The main rails are installed side by side to the roof of the SF so that the working spaces of the OMs overlap.
5. *Factory disassembly system.* The SF and subsystems need to be dismantled after completion of the assembly work. Therefore, two jib cranes are installed on the roof of the SF (note: after completion the roof of the SF becomes part of the building and therefore does not have to be disassembled).
6. *Welding robot system for columns.* The welding robots consist of two robots travelling on a steel ring that is temporarily attached to the columns. Before the

welding operation the columns are aligned by the alignment and accuracy measurement system (see later).

7. *Alignment and accuracy measurement system.* In the MCCS, the alignment and displacement of each column installed by the manipulators are measured by 1D laser systems and reflector sheets installed on each column.
8. *Real-time monitoring and management system.* The MCCS has a control room located under the roof of the SF, in which the activities of OMs, CS, logistics and delivery systems, welding robots and alignment and measuring system are monitored and supervised. All in all the MCCS integrates the following management and real-time control capabilities:
   - *Machine operation management system.* This system visualizes for the human site supervisors the status of CS, HDS, and the OMs, as well as the conditions (for example the temperature) in and around the factory.
   - *Progress management system.* This system visualizes for the human site supervisors in real time 3D information on the materials processed by HDS and OMs.

End-effectors Compared to other systems, the manipulators are equipped with relatively simple (hook) end-effectors for material delivery and component installation. However, the column catcher that is part of the CS can be considered as an advanced end-effector that cannot be found in this form in other systems.

System Variations The system can be varied in length and width on the basis of a grid system that synchronizes building systems and manufacturing systems. As the steel stilts (part of the CS) holding the SF rest on the outer rows of columns of the two front sides of the building, and the factory roof structure spans the building without additional support, the distance allowed between the two face sides is limited and allows only for small changes.

Robot-Oriented Design On a component level, the column components in particular, as well as the facade elements, are designed according to ROD principles. However, compared to other systems, no compliant joints or plug and play joints are used. On the contrary, devices such as the column catcher are used, so that actually no major system specific redesign of the building components or joints is necessary for operation of CS and SF. On a building level, the structure of the SF and the building's columns grid have to be adjusted to each other. From an urban standpoint, the location of the intake ports through which the OMs directly lift components can be considered in terms of ROD (e.g., it would be advantageous if the orientation of the building allows trucks to access this side of the building easily during construction).

Efficiency Analysis

The MCCS produces steel-based buildings through an SF that operates on top of the building. The main elements of the system are two gantry-like automatic/robotic OMs that are fixed to the roof of the SF and the automated CS. The OM system of the MCCS is lighter than, for example, the OM system of FACES and thus also allows for faster operation. The two OMs can, through an opening in the factory on the side, pick up elements and install them directly.

Floor Erection Cycle Maeda operates its MCCS on a 9-day FEC. This FEC necessitates parallel work on three levels.

Technical Data Speed of Equipment The CS and the overhead OMs are central to the system. The lifting of the whole factory to the next level can be accomplished by the CS at a speed of 0.37 m/min and the lifting of the SF's stilts (necessary to the placement of a new steel column under each stilt) at a speed of 2.45 m/min. The OM system and its individual links/components can be operated at a speed of 20–25 m/min.

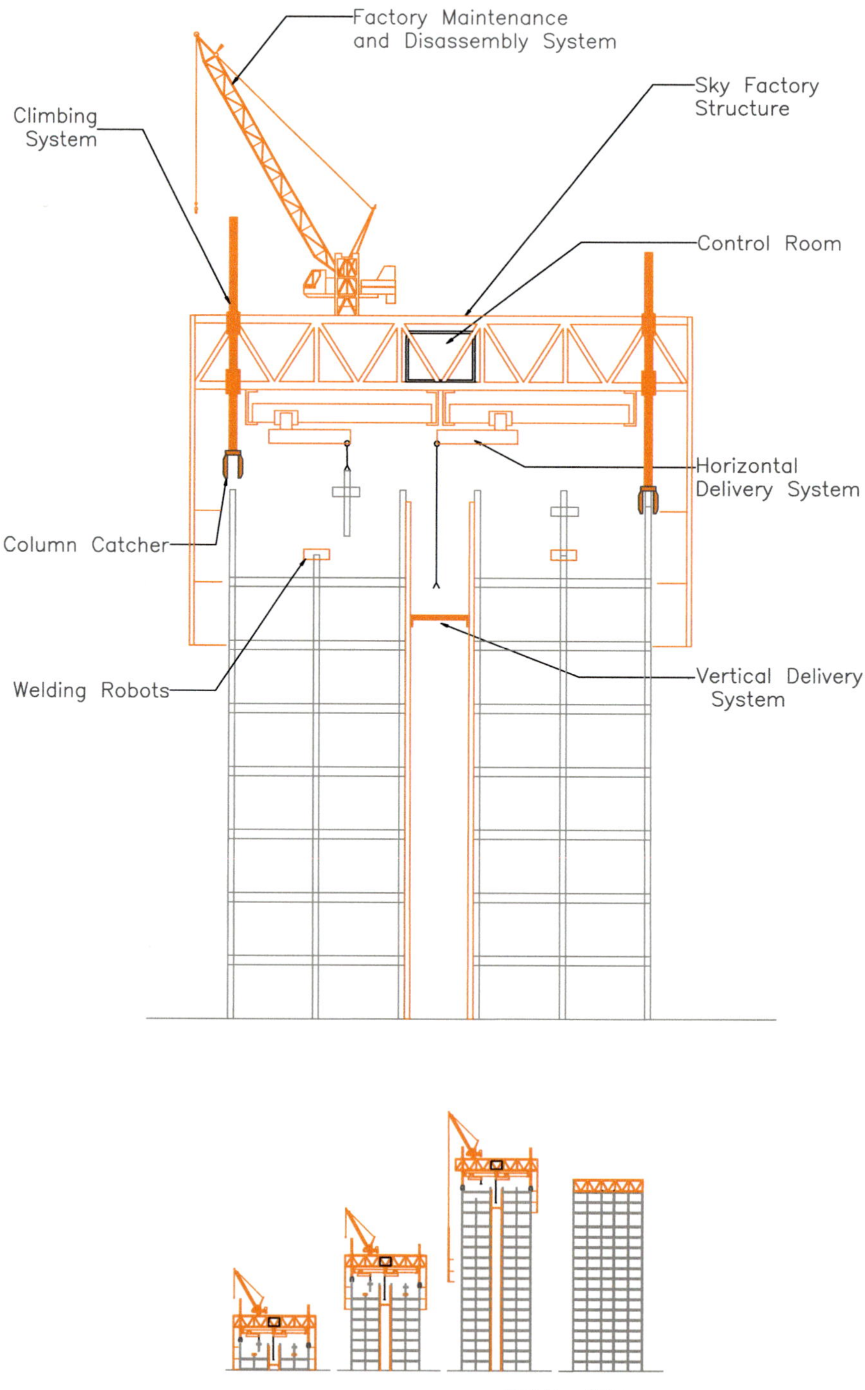

Mast Climbing Construction System (MCCS), Company: Maeda Corporation, Japan

Figure 2.96. Sky factory, exterior view.

Figure 2.97. Sky factory, exterior view.

Figure 2.98. Sky factory, exterior view.

Figure 2.99. Each stilt is equipped with a column catcher.

Figure 2.100. Rail-guided overhead manipulators.

Figure 2.101. Overhead manipulator, detailed view.

Figure 2.102. Rail system, detailed view.

Figure 2.103. Sky factory, side view.

Figure 2.104. Sky factory, front view

Figure 2.105. View on climbing mechanisms actuating the stilts.

Figure 2.106. View on climbing mechanisms actuating the stilts.

Figure 2.107. View on climbing mechanisms actuating the stilts.

2.1.5 Roof Push-up Construction Method

Company: Takenaka Corporation, Japan
Category: SF (moving upwards) supported by building

Analysis of Technology and Composition

Evolution Scheme The Roof Push-up system concentrates manufacturing activity in an SF (Figure 2.108) where low- and high-level components such as steel columns, steel beams, floor elements, and plumbing units are assembled in a weatherproof and SE. At first, on the ground, SF, OMs, and the material lifting equipment are installed. The factory itself is set up at a ground level by use of a jib crane. Then, step by step, one floor after the other is produced inside the SF in an SE. Stilts, which are part of a CS, rest the SF on top of columns erected and perform the climbing process.

Elevation Special features of the system are the integrated jib crane on top of the system, which is used to lift up and position prefabricated plumbing components and units, as well as a method/equipment that allows the convenient overhead installation of plumbing interior finishing. Furthermore, as in most other systems, the installation of components of the load-bearing structure is done by rail-guided OMs. An automated VDS over a shaft within the floor area delivers material to the SF. A weatherproof SF surrounds the floor being produced and extends over the floor below. The FEC is timed so that the production of the building's structure on the operational floor (*n*th floor) is completed at the same time as the installation of facades and additional room facilities on the floor below the *n*th floor ($n - 1$ floor).

Figure 2.108. Sky factory, exterior view.

Ground Plan Components from the ground floor are lifted by the VDS and then delivered to the *n*th floor where the components are delivered to the desired position within the *n*th floor by the OMs. The OMs can move horizontally and can be controlled remotely to position columns and other steel components. The SF is raised by raising the CS's steel stilts and placing new columns under them. When the stilts supporting the roof are raised the new column is positioned, aligned, and fixed below the raised stilt.

Subsystems The Roof Push-up system consists of a set of integrated subsystems:

1. *Vertical delivery system (automatic lift).* An automatic lift running in a shaft within the floor serves as VDS and delivers parts and components into the SF. After completion, the shaft can be closed or used as a vertical shaft for stairs or elevators.
2. *Sky factory structure.* The SF structure covers the building and provides a weatherproof SE (Figures 2.111 and 2.113). It integrates the jib crane on top, the manipulators, the lifting mechanism, the mechanism for lifting floor elements after overhead work (overhead interior finishing system), and working platforms for exterior facade work. After completion, the roof structure of the factory is integrated into the building as the roof or at least as one of the uppermost floors.
3. *Climbing system.* Hydraulic actuators are used to lift up the floor step by step (Figure 2.114). The lifting operation is carried out in several steps. First, the climbing stilts supporting the SF structure are lifted up by hydraulic jacks. Second, a temporary column is installed and the structural building columns are positioned, aligned, and fixed. Third, the stilts supporting the SF are lowered again and fixed onto the installed column; the temporary column is removed. This is done for each column on each floor. Finally, the factory can be raised.
4. *Sky factory horizontal delivery system.* Two lengthwise rail-guided OMs (Figures 2.112 and 2.115) position components delivered into the factory by the VDS. The working areas of the two OMs overlap in the area of the central row of columns. Each manipulator has 5 DOFs (starting from base point: translation lengthwise, translation right/left, rotation parallel to floor level, rotation perpendicular to floor level, travelling of hoist along final link axis). Each manipulator's final link integrates an end-effector that can embrace the lifted climbing stilt and position column elements under it.
5. *Facade element positioning system (optional).* Along the corners of the building, rail-guided hoist-like manipulators (3 DOFs: lift/lower, move along building corner, move to building/move away from building) can travel and pick up, lift, and position facade elements.
6. *Steel column welding system (optional).* After positioning and alignment of column elements, optional welding robots (rail-guided, attached to the column) can be used.
7. *Singe task construction robots (optional).* STCRs for concrete placement and levelling can be used within the SF (Figure 2.116).
8. *Plumbing unit installation system.* A jib crane on top of the factory lifts and positions prefabricated plumbing units once a floor's main structure has been erected by the OMs. An intake port on top of the SF can be opened and closed for this purpose.

9. *Overhead interior finishing system.* To further reduce work in elevated places, overhead interior finishing (including plumbing and the installation of aeration systems) for floor components is done from the finished lower floor ($n-1$ floor). The thus on the site prefabricated floor component is then raised and installed on the uppermost floor. As such, no temporary scaffolding is required.
10. *On-site control room.* An on-site control room for supervising, coordinating, and (remotely) controlling all work activity is integrated into the SF structure above the operation level of the OMs.

End-effectors The manipulators' 5 DOFs and double fork structure allows them to perform all necessary installation operations. As end-effectors, simple hooks and automated and nonautomated clamps are used (e.g., for transporting and positioning of columns and beams by OMs).

System Variations The system has so far been applied without the optional subsystems mentioned earlier. The optional subsystems are subject to R&D and are conceived to be integrated with the other subsystems. Furthermore, as Takenaka had also developed previously a series of STCRs (e.g., for concrete distribution, floor finishing; see **Volume 3**), the integration of those with the Roof Push-up system would add additional optional capabilities and thus variety to the system. The system in total already has inbuilt flexibility, allowing (from ground plan view) nonrectangular buildings to be manufactured.

Robot-Oriented Design On a component level the design of the column elements is a central element of ROD. The columns already contain beam elements welded to them on various sides. This allows the components to be easily transported by the OMs' double forks. Furthermore, it allows that the positioning and adjustment of beam components between the formerly mentioned components is afterwards simplified. On a building level, and in terms of ROD, a floor plan that concentrates on the one hand on the functional elements (plumbing, vertical ducts, elevators, stair cases, etc.) and on the other hand on the office rooms, allows for the efficient use of the plumbing unit installation system – a speciality of the Roof Push-up system.

Efficiency Analysis

Takenaka's Roof Push-up operates an SF on top of a building. In the SF, overhead manipulators assist with the installation of components. One specialty of the system is the full integration of a jib crane on top of the SF, which assists in delivering prefabricated plumbing elements. Another specialty is the use of a system that supports the installation of plumbing elements under the ceiling elements.

Floor Erection Cycle Takenaka operates its system on a five-and-a-half day FEC (Figure 2.109). Similar to other companies, Takenaka can complete a cycle working on several floors in parallel (construction floor, carrying floor, and interior finishing in lower floors). The specialty of the system is the carrying floor.

Experiments with Degree of Automation/System Configuration/Worker Teams To optimize the sequence of component installation and the arrangement of labour, Takenaka started to develop a system to simulate and optimize the building assembly

Table 2.3. *Simulation of processes to optimize performance cross all construction steps*

Step 1: Design	Step 2: Construction planning	Step 3: Construction management
• Process planning • Planning the arrangement of labour • Planning the method of material carriage • Planning the arrangement of subsystems	• Realizing work design • Best sequence of material instalment • Best arrangement of labour	• Construction site monitoring • Accuracy management • Safety management • Material management • Labour management

Adapted from Morita et al. (1993); Fujii et al. (1995).

process, executed by the Roof Push-up system (Table 2.3). The system was intended to generate and optimize the following data sets (Morita et al., 1993; Fujii et al., 1995) necessary to set up an optimized workflow in the on-site factory for a given building configuration:

- Number of construction processes
- Number of workers/worker teams
- Number and location of temporary storage
- Sequence and interval of component pickup and installation

The R&D vision for that system was that it would eventually automatically generate the detailed design (optimized for Roof Push-up method in the sense of ROD), the construction planning/sequence generation, and that on the basis of the generated data construction execution and monitoring could be automated. In the end, only the software module for construction planning/sequence optimization was realized.

Using the realized construction planning simulation/optimization system, Takenaka, in the first projects using the Roof Push-up system, was able to determine, for example, the optimal number of workers or teams of workers (Morita et al. 1993; Fujii et al., 1995).

Productivity The following figures relate the five-and-a-half day floor cycle execution scheme to the number of workers and types of workers needed per day (Figure 2.110):

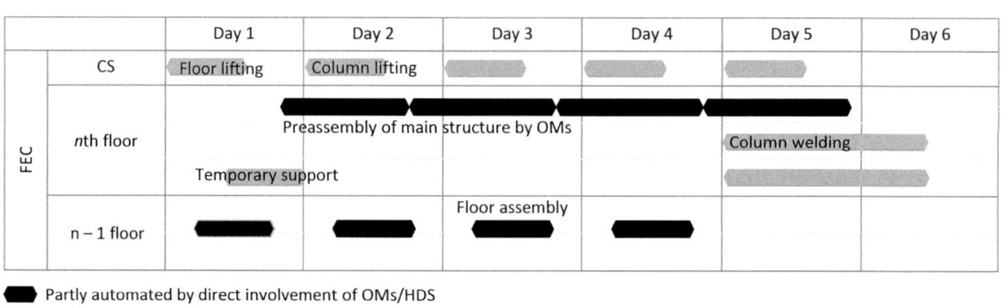

Figure 2.109. Floor erection cycle (adopted from Morita et al., 1993; Fujii et al., 1995).

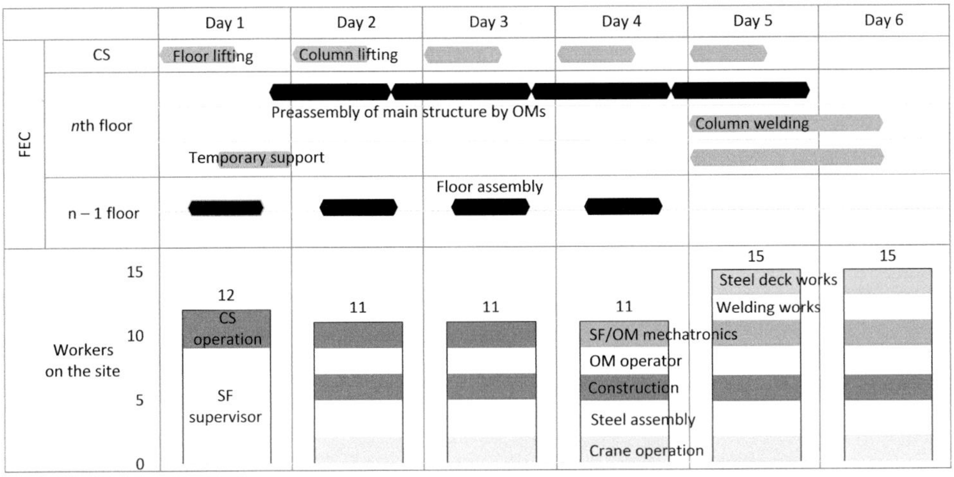

Figure 2.110. Interrelation of FEC and labour resources required (adopted from Morita et al., 1993; Fujii et al., 1995).

- Day 1: about 12 workers (CS jack operators, supervision workers)
- Days 2–4: about 11 workers (steel workers, erectors, OM operators, precision operation workers)
- Days 5 and 6: about 15 workers (steel deck workers, welding workers, steel workers, erectors, OM operators, precision operation workers)

Results The "work unit ratio" in the context of the number of construction activities conducted was 0.25 men/m^2. The value achieved is close to the average value usually achieved in (conventional) steel frame construction projects. This result can be expressed also in FECs. It takes almost 6 days to complete a floor with a "work unit ratio" of 0.17 men/m^2 in a standard project (Morita et al., 1993; Fujii et al., 1995). There is, therefore, room for improvement regarding the overall productivity of the Roof Push-up system.

Product Quality and Process Monitoring Takenaka supervises the construction progress in real time from a set of computers. It monitors working operations through video cameras and collects data about the precise installation of the components by a set of sensors installed in the Roof Push-up factory. The continuous monitoring allows Takenaka to simplify the quality assurance process as well as to collect data about improvement of work processes and work productivity.

Central Control System Operations in the Roof Push-up method are controlled by an RTMMS operated from a central control room on-site. The system consists of three subsystems. The operation monitor subsystem monitors work-in-progress throughout the whole construction site. The data from this system are used to inform workers about the progress of construction and their individual work tasks. The communication subsystem is used to give direct instructions to workers working on the movable floor or floor below. The precision monitor subsystem can provide a real-time monitor on floor and column installation.

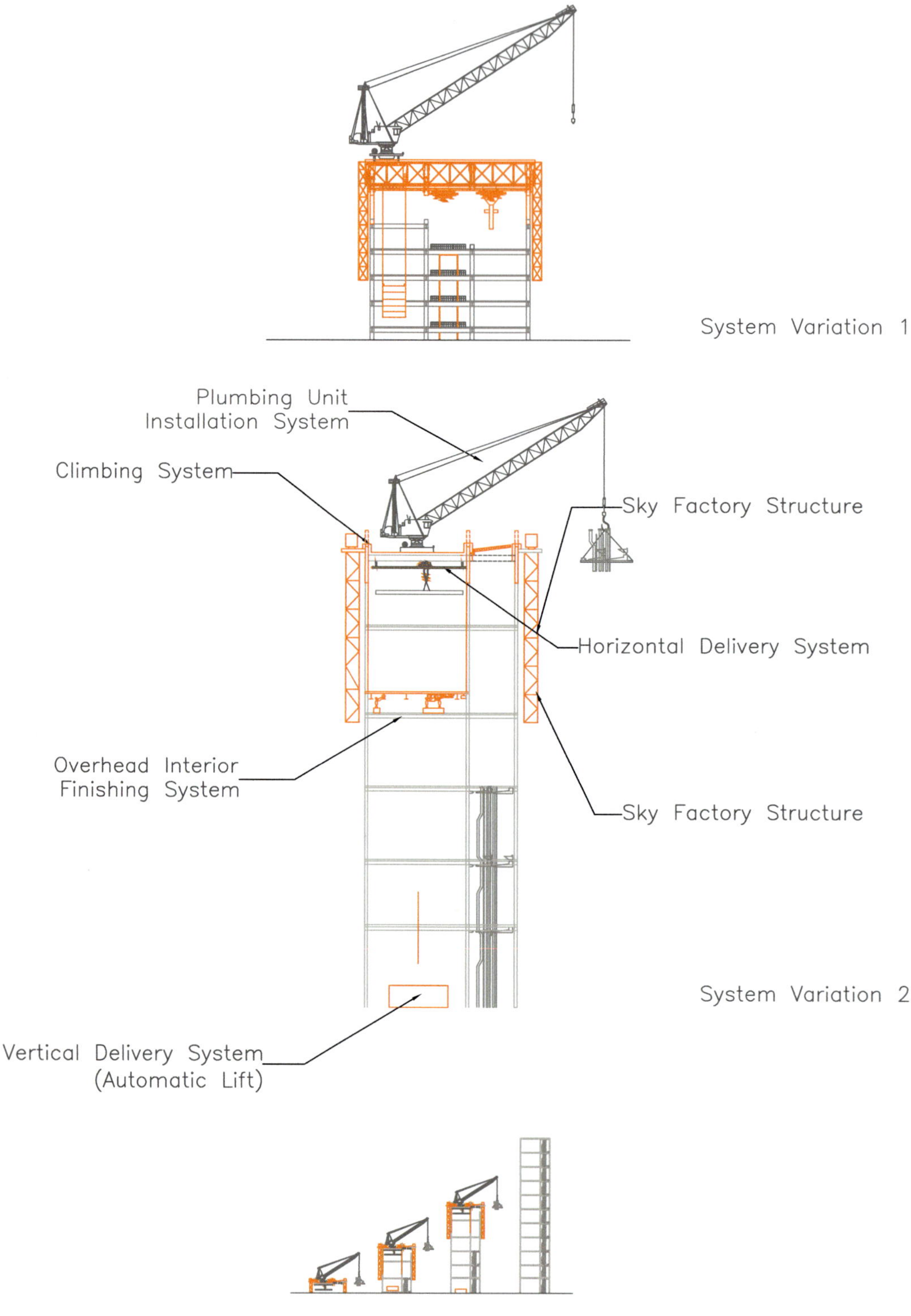

Roof Push-up Construction Method, Company: Takenaka Corporation, Japan

Figure 2.111. Sky factory, interior view.

Figure 2.112. Overhead manipulator installing a floor component.

Figure 2.113. Sky factory, interior view.

Figure 2.114. Climbing mechanism actuating the stilts.

Figure 2.115. Experiments with a variation of an overhead manipulator.

Figure 2.116. Single-task construction robot used for concrete distribution. (Image: Takenaka)

2.1.6 Roof-Robo Automated Construction System

Company: Toda Corporation, Japan
Category: SF (moving upwards) supported by building

Analysis of Technology and Composition

Evolution Scheme The SF (Figure 2.117) is set up on the ground and then lifted up along the stilts of the CS supporting it. The first three floors are assembled in a conventional way. From the fourth floor upwards the full operation of the system starts including assembly of superstructure columns on the *n*th floor, plus parallel processing starts on floors $n-1$, $n-2$, and $n-3$. The SFs' roof truss structures can be integrated into the building as the basic steel structure for the uppermost floors or roof of the building.

Elevation during full operation, the SF covers the uppermost four floors and ensures that all processes performed before (and during) the attachment and sealing of the

Figure 2.117. View into the sky factory and operational levels. (Image: Hasegawa, 1999: p. 467)

facade are done in an SE. OMs are attached to the roof of the SF and are able to pick up components directly from the VDSs. On the uppermost floor under the roof (*n*th floor) the OMs position the superstructure columns, and the automatic alignment system is applied. On lower levels, beams and, thereafter, the facade elements are installed. Furthermore, after installation of the facade, work on the fireproof coating and interior finishing starts.

Ground Plan The ground plan shows the arrangement of the SF as well as the position of the GFs. The GFs are connected to the SF directly via two automatic VDSs to be able to establish JIT and JIS components (to-site) delivery and installation on-site process chains.

Subsystems The Roof-Robo system consists of a set of integrated subsystems:

1. *Sky factory structure.* The SF structure encloses the workspace of the top three or four levels, and thus creates a weatherproof, safe, and structured environment for all assembly tasks. Once the facade is installed and sealed, the building leaves the SF and further final finishing work can be done in the sealed interior. The factory roof's steel truss structure can be integrated into the building as the final floor.
2. *Climbing system.* The CS is hydraulically actuated and controlled automatically. The steel stilts supporting the SF are pushed up by the actuation mechanism located in the roof truss structure. The stilts rest on the superstructure elements of the building located in its four corners. Each corner of the superstructure consists of four columns, which are positioned, aligned, and fixed around the stilt supporting the SF structure. Once they are fixed and connected by beams, the stilt is lifted and fixed by bridging the new superstructure segment. This is done in all four corners of the building; then the factory can be lifted up along the stilts and fixed into the superstructure.
3. *Vertical delivery system.* An automated lift transports structural elements into the SF. The lift is especially designed to transport and orientate the superstructure components – the main element of the building – to the OMs. By following a curve on lower level, the lift raises the superstructure columns up from a horizontal position (which is convenient for handling on the ground) to an upright position, allowing them to be directly picked up by the OMs.
4. *Vertical/horizontal finishing material delivery system.* The finishing material is delivered from the GF to the SF and to the installation location by a forklift-like mobile robot. The mobile robot is compact and lightweight and can thus travel vertically by an automatic lift, with which it can communicate. The mobile robot has forks and a multipurpose end-effector that allows the transport of palletized and nonpalletized material.
5. *Horizontal delivery system.* Two OMs with 4 DOFs each (translation along the *x*-axis, translation along the *y*-axis, rotation, and lower/lift of end link along the *y*-axis; payload: up to 6 tons; operating speed: up to 35 m/min; Figure 2.118) are operated on a rail system attached to the SF's roof. They are used to position the building's superstructure components, beam/girder elements connecting the four superstructure trusses of the building, and floor slab elements.

6. *Curtain wall handling system.* Within the SF, but on the levels under the uppermost operational *n*th floor (so on $n - X$ floors), Toda uses its curtain wall installation single-task construction robots (STCRs) to install prefabricated facade components. The elements are transported by the vertical/horizontal finishing material delivery system to, for example, the $n - 1$ floor and are then lowered by the system to $n - 2$ level and positioned (see also **Volume 3**).
7. *Fireproof coating system.* On the lower levels within the SF (*n*th floor to $n - 3$), or even on the lower and already sealed floors under the SF, STCRs are utilized for placing a fireproof coating onto the assembled steel components. The STCRs can be repositioned and moved on an individual floor level horizontally or transported vertically by the vertical/horizontal finishing material delivery system.
8. *Ground factory (optional).* Toda can install GFs for material handling, material storage, and on-site component prefabrication. The GFs are additional on-site factory units that are connected to the vertical transportation systems. They allow trucks delivering material to drive into the GFs and logistics and installation preparation in an SE.
9. *Automatic column alignment and accuracy measurement system.* After positioning by the OMs, the columns of the superstructure are temporarily fixed by automated clamps. The alignment of the column over component integrated reference points is measured by 1D laser scanners. Then the clamps automatically align the column to the required position.
10. *Work process and material delivery management system.* Toda uses a software tool to simulate and visualize work processes and to automatically generate work instructions and material delivery schedules, and to manage and control the JIT material flow and the operation of the two VDSs in real time.

End-effectors Both overhead gantry type manipulators and the V/HDS use simple (conventional crane-like) multipurpose end-effectors.

System Variations In its current version the system can be applied to buildings with different heights, widths, and lengths that are based on superstructures. The CS must be able to climb along with the superstructure. The length and width of the building are restricted by the strength of the SF's roof structure (and thus its ability to span space between the superstructures).

Robot-Oriented Design On a component level, ROD is applied to the column components used to build up the superstructure and to the facade elements. The columns have temporarily fixed steel plates attached at the top end that allow them to be picked up by the manipulator's hooks. Furthermore, the components have integrated reference points that allow the efficient use of the automatic column AAMS. On a building level, it has to be considered that for efficient operation, the building structure needs to be based on a superstructure. Unlike other systems, the roof structure system is supported by only four stilts, which rest on and climb along the four corners of the superstructure. The reduction of factory structure supporting stilts reduces the complexity of the CS. On an urban level, the allocation of the vertical lifts and the ground-level adjunct GFs or MHSPYs need to be considered when deciding on the orientation of the building in the environment.

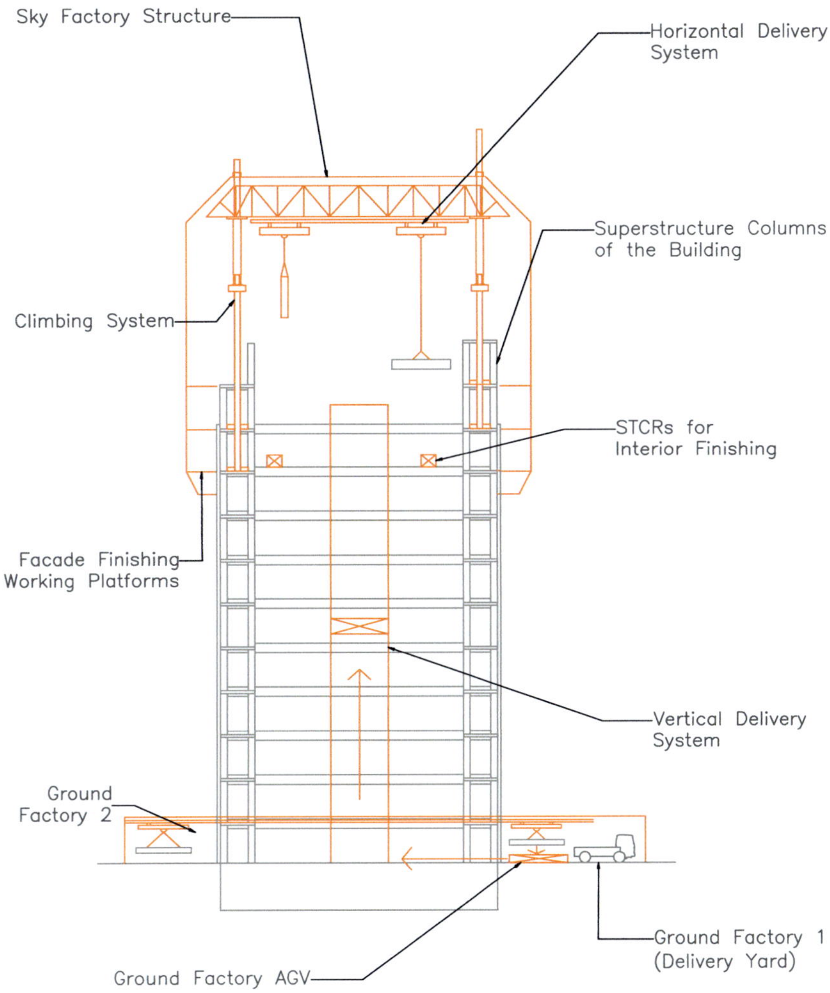

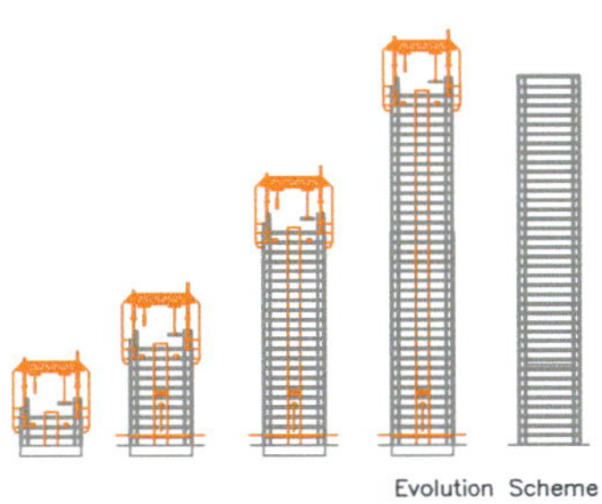

Roof-Robo Automated Construction System, Company: Toda Corporation, Japan

Figure 2.118. Interior view of an SF with an OM in action. (Image: Hasegawa, 1999: p. 468)

2.1.7 Shimizu Manufacturing System by Advanced Robot Technology (SMART)

Company: Shimizu Corporation, Japan
Category: SF (moving upwards) supported by building

Analysis of Technology and Composition

Evolution Scheme The SMART SF roof structure (and the subsystems attached to it) is assembled on the ground and then lifted along the supporting steel stilts up to a height at which operation can start. To begin with, the supporting stilts rest on levels that have already been finished. After the construction of the first floors by the system, the stilts/climbing mechanism can use these newly constructed floors to lift

Figure 2.119. Sky factory, exterior view.

the SF. Full utilization of the SF (including its capability of also providing a structured work environment for the outside facade work) can start from the third floor upwards (Figures 2.140 and 2.141). Furthermore, after the construction of the fourth floor, a material handling system can be deployed on the ground floor. After the dismantling of the factory covers and manipulators (Figures 2.137–2.139), the roof structure of the SF can be integrated into the building as the uppermost floor (e.g., as the service floor containing the aeration system, etc.).

Elevation The SMART concentrates work activity in an SF (Figure 2.119). In contrast to other systems, the OMs do not have a fixed main trajectory (such as one to three lengthwise trajectories as is the case of ABCS, for example) but can travel freely throughout the whole system. Up to 25 manipulators (trolley hoists) can operate in parallel, and travel from the material handling space on the ground floor, via an automatic lift into the SF where they can move to any component installation location via a (movable) gantry crane rail system. Furthermore, in contrast to other systems the steel stilts that support the SF and are part of the climbing mechanism do not rest on top of the columns but between them – thus the climbing process interferes less with component assembly process. Within the SF, work is done on three to four levels in parallel. Components are positioned and aligned on the uppermost floor (the *n*th floor). Activities such as component fixation, interior finishing, and facade work (over platforms provided by the SF) are done on the floors below. Once the SF moves a floor higher, a completely finished floor leaves the factory at its lower end.

Ground Plan A system of mobile overhead cranes is synchronized with the building shape and the column grid system of the building. Multiple mobile overhead cranes travel the building in a fixed lengthwise direction. However, the trolley hoist manipulators can travel perpendicular to the direction of the cranes (and therefore also between them). Thus the trolley hoist manipulators can travel to any point in the *x/y* (length/width) direction. As the stilts supporting the SF structure do not rest on columns but between columns, they do not permit the trolley hoists to position components directly over a column. The trolley hoists pick up components from the ground floor, travel with the component over the automatic vertical lift into the factory, and then travel via a transport channel (multiple mobile overhead gantry cranes can form a line profile), allowing the trolley manipulators to move along it.

Subsystems The SMART system manufacturing the building on-site by way of an SF consists of a multitude of integrated subsystems. The trolley hoist manipulators freely travel throughout the system by automatic vertical lifts and automatic overhead cranes. They thus integrate vertical and HDSs and take away the necessity of handing over components between them, thus eliminating waste processes. (Consider the waste that occurs in other systems as a result of the handover between vertical and HDSs, e.g., the resulting idle time of the component and the time necessary to accomplish handover; in addition consider the possibility of damage to components during handover.)

1. *Material handling system (ground level).* As the trolley manipulators pick up material from the ground level, a material handling space on the ground is necessary that allows an overhead crane or rail system to be positioned directly

over the components to be stored or delivered. Picking up directly from a truck that can be driven into the material handling space allows for JIT processing of components delivered to the site. The material handling space can be constructed as a temporary structure next to the building itself, or integrated into the lower floors.

2. *Sky factory structure*. The SF (Figures 2.130–2.133 and 2.136) completely covers the building and creates a weatherproofed and structured working environment for all automatic and manual work activities conducted between the *n*th floor and floor $n - 3$. The SF roof structure is supported by steel stilts that rest between columns on the floors under the uppermost operational floor. The roof structure cantilevers over the building's edge and thus also provides working platforms for the facade work. The vertical trolley manipulator lifting system, as well as the system of overhead cranes (which provide the basis for the free travelling of trolley manipulators in the system), are directly integrated and mounted onto the factory roof structure.
3. *Climbing system*. Four lifting stilts are positioned inside the factory to lift the whole factory (Figure 2.135). In contrast to many other systems that use SFs, the hydraulic lifting mechanism is not located in the roof structure itself but on the bottom of the stilts. Inchworm-like, the mechanism moves from floor to floor, pushing up the stilts and thus the factory.
4. *Trolley hoist manipulator system*. In the SMART, up to 25 trolley manipulators are operated in parallel (Figures 2.142–2.144). Each trolley manipulator can travel between material handling systems on the ground and the SF's automatic overhead gantry crane system, thus integrating pick-up, vertical delivery, and positioning processes.
5. *Sky factory horizontal delivery system*. Multiple mobile gantry-type overhead cranes travel the building in a fixed lengthwise direction (Figure 2.145). However, the trolley hoist manipulators can travel between the cranes (Figures 2.146–2.148). Thus, the trolley hoist manipulators can travel to any point on the *xy*-axis. Multiple overhead cranes can build a line (transport channel) along which the trolleys can travel.
6. *Vertical delivery system (lift of trolleys/manipulators)*. The trolley hoists pick up components from the ground floor and travel with the component via the automatic vertical lift into the SF (Figures 2.149–2.153). Together with the automatic overhead crane system, the trolley manipulators used during a component positioning operation in the SF have 4 DOFs (translation of overhead crane, travelling of trolley manipulator along crane profile, turning of trolley head, lifting/lowering). When the lifting process and the travelling of the trolley in the material handling system on the ground are taken into account, the trolley manipulators actually have more than 4 DOFs.
7. *Welding robot system*. Compared to other welding robot systems, the Shimizu welding robot system used within the SMART is compact and lightweight. It consists of a ring that encircles the column and a single welding robot guided by this ring. Many of these systems can be used in parallel. Each welding robot system determines its welding path by a laser sensor and automatically welds the steel columns together. Eventually the transportation device still attached to the welded column is removed automatically.

8. *Real-time monitoring and management system.* Each construction element is tagged with a barcode. This code is scanned at certain locations during transport by on-site logistics and positioning systems and the individual elements are then processed by the distribution systems according element-individual work instructions. For an accurate positioning of the steel columns, a laser measurement system is used. A control room for supervising and monitoring the manufacturing process is located in the roof structure of the SF.

End-effectors All trolley hoist manipulators are equipped with multipurpose hooks, which are lowered/lifted by two ropes from the trolley manipulator's main mechanical part. The simplicity of the end-effectors is accomplished as the system does not necessitate positioning columns under the SF's supporting stilt and as the trolleys pick up components already in the correct orientation (e.g., columns in upright position) in the ground floor.

System Variations The SMART is a fully modular kit (Figures 2.154–2.158). The system can be used to construct rectangular as well as nonrectangular buildings with an orthogonal alignment of components. If nonrectangular buildings are built, the shape of the building has to be synchronized with the length of the overhead cranes, providing the travelling axis for the trolley manipulators. The SMART can be operated with one, two, or even more vertical trolley transport channels. The more trolleys that are intended to be used in parallel, the more transport channels are needed.

Robot-Oriented Design On a component level, the complexity of the column installation process is reduced by ROD through the use of compliant and self-adjusting joints (rack and pinion system; Figure 2.134). On a building level, the structure of the overhead crane system has to be synchronized with the shape and dimensions of the building, and the shape and dimension of the SF structure. On an urban level, according to ROD principles the location of the material handling system on the ground and the location of the vertical transportation channels ought to be considered when designing and deciding on the orientating of the building.

Efficiency Analysis

Shimizu's SMART erects steel-based buildings. The factory on top of the building contains a system of gantry OMs that allow robotic trolley hoists to freely travel in the systems and thus to transport and position elements (columns, beams, floor slabs, facade elements).

Detailed Project Schedule An analysis comparing conventional construction and SMART-based construction reveal that the SMART system reaches the breakeven point by about the tenth floor (Maeda, 1994; Maeda & Miyatake, 1997). The SMART is slower at the beginning because of the time required for construction planning, system programming, and system set-up. For a 20-storey building the SMART is able to reduce the time necessary to construct the building by about 15% (Figure 2.120). If the building is higher than 20 storeys, higher gains in reducing the construction period are possible.

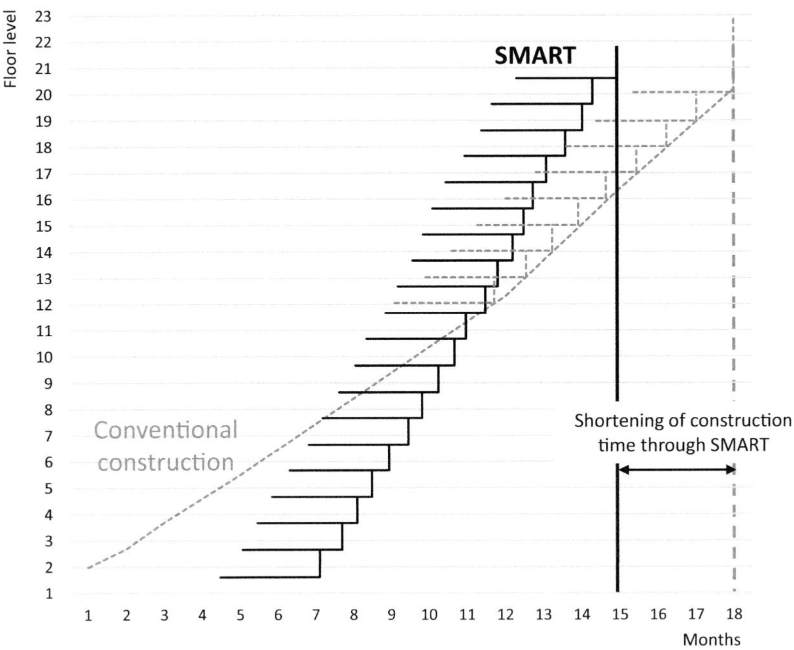

Figure 2.120. Shortening of construction period: comparison between conventional and SMART system of construction (adopted from documentary material received from Prof. Maeda/Shimizu; see also Maeda, 1994; Maeda & Miyatake, 1997).

Floor Erection Cycle Shimizu starts to operate the SMART on a 7-day FEC in the first projects, and has managed to reduce cycle time to 5.5 days after a certain number of projects and then down to 5 days after a further number of projects (Figures 2.121–2.122). The cycle time can be improved as a result of learning effects in operating the equipment (see also below *Learning Effects*).

Technical Data, Speed of Equipment In a conventional tower crane–based construction process, components can be delivered only sequentially, one after another. The SMART, with its multiple robotic trolley hoists (up to 25), allows the parallel operation of multiple pick-up and positioning operations (Figures 2.123 and 2.124).

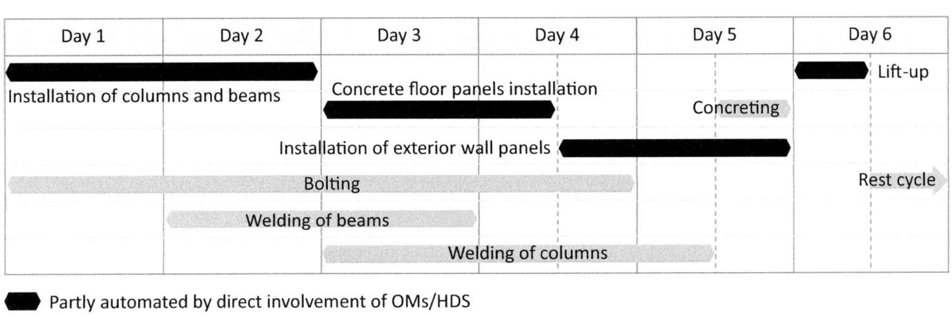

Figure 2.121. SMART applied with a six day FEC process (adopted from documentary material received from Prof. Maeda/Shimizu; see also Maeda, 1994; Maeda & Miyatake, 1997).

Table 2.4. *Comparison of Yokohama project with Nagoya project*

	Rail city Yokohama building	Nagoya Juroku bank building
Total floor area	75,000 m	20,000 m
Standard floor area	2,100 m	750 m
Building height	132 m	88 m
Number of storeys (above ground)	30 storeys	20 storeys
Shape of standard floor	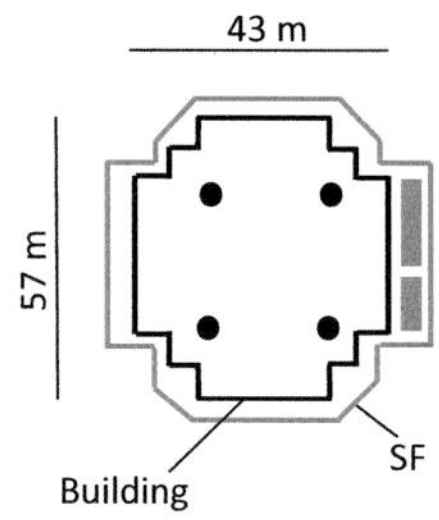	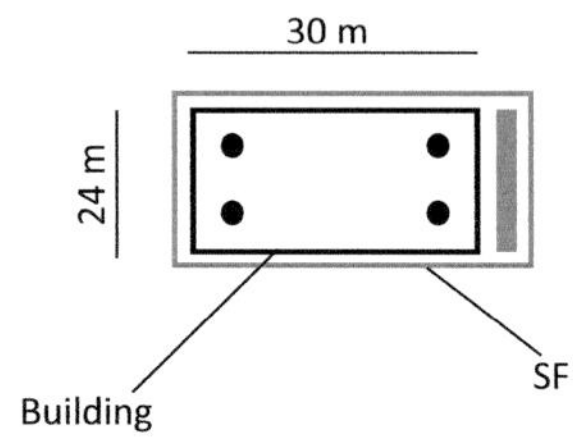
VDS	2 sets	1 sets
Overhead crane	24 sets	10 sets
Trolley hoist	10 sets	5 sets
CS stilts	4 sets	4 sets
Hydraulic jack	150 ton × 12 sets	120 ton × 12 sets

Adapted from Maeda (1994); Maeda and Miyatake (1997).

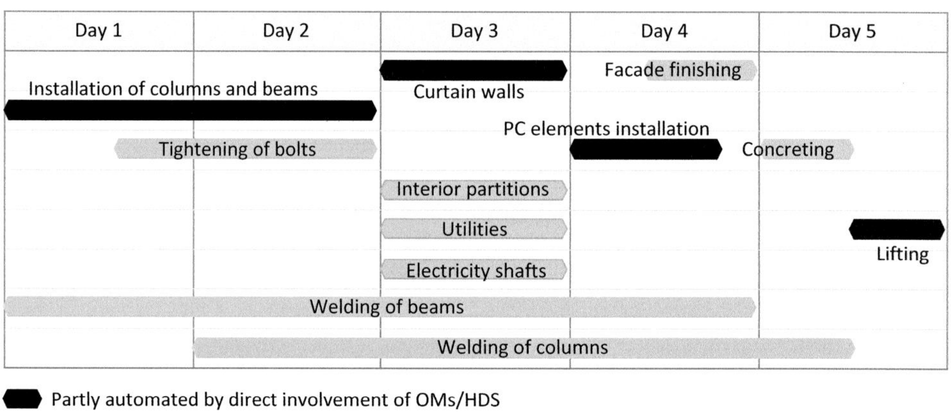

Figure 2.122. SMART applied with a five-day FEC process (adopted from documentary material received from Prof. Maeda/Shimizu; see also Maeda, 1994; Maeda & Miyatake, 1997).

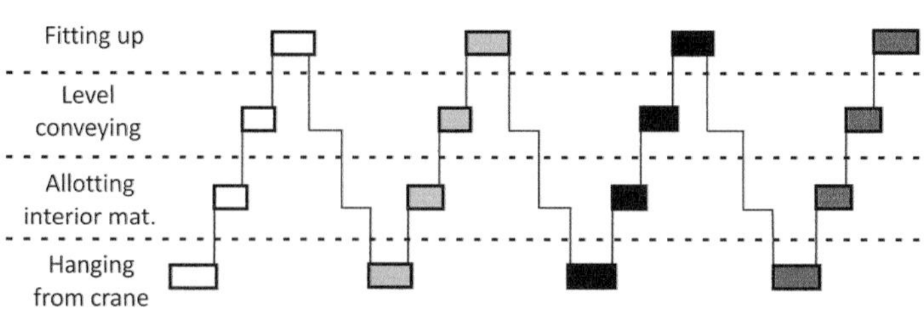

Figure 2.123. Component positioning/installation by conventional tower crane (adopted from documentary material received from Prof. Maeda/Shimizu; see also Maeda, 1994; Maeda & Miyatake, 1997).

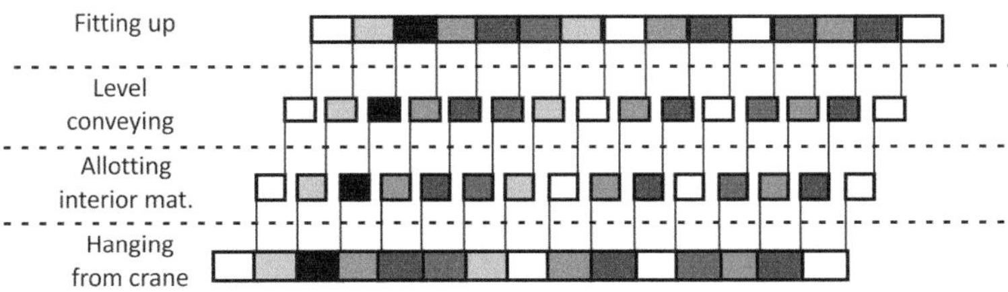

Figure 2.124. SMART's component positioning/installation strategy (adopted from documentary material received from Prof. Maeda/Shimizu; see also Maeda, 1994; Maeda & Miyatake, 1997).

Experiments with the Degree of Automation/System Configuration/Worker Teams Similar to Obayashi's ABCS, or Big Canopy, the gantry OMs and trolleys can be operated manually or automatically. The most practical solution is for the gantry OMs and trolleys to operate automatically during transportation to the installation location and assembly, and positioning to be done manually. In various projects Shimizu experimented with different configurations of the site factory, and thus with different amounts and combinations of gantry OMs and trolley hoists. Table 2.4 compares exemplarily two projects with substantially different configurations.

Productivity SMART enables improved labour productivity. The on-site productivity (measured in man-hours) needed to complete the various working tasks in the first applications of SMART was increased by 50% compared to conventional construction methods (Figures 2.125 and 2.126). The highest rate of improvement rate was to be found with the installation of the facade.

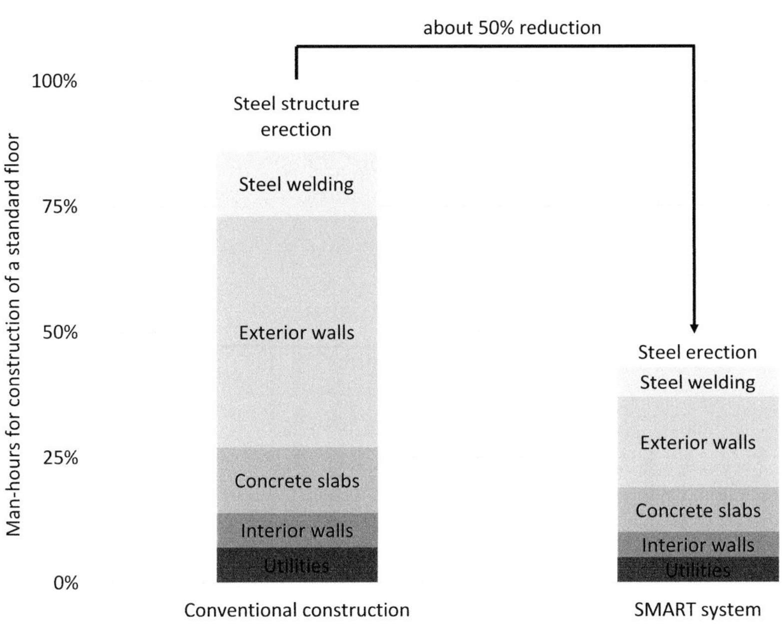

Figure 2.125. Reduction of man-hours related to the construction of a standard floor manufactured by the system in full operation (adopted from documentary material received from Prof. Maeda/Shimizu; see also Maeda, 1994; Maeda & Miyatake, 1997).

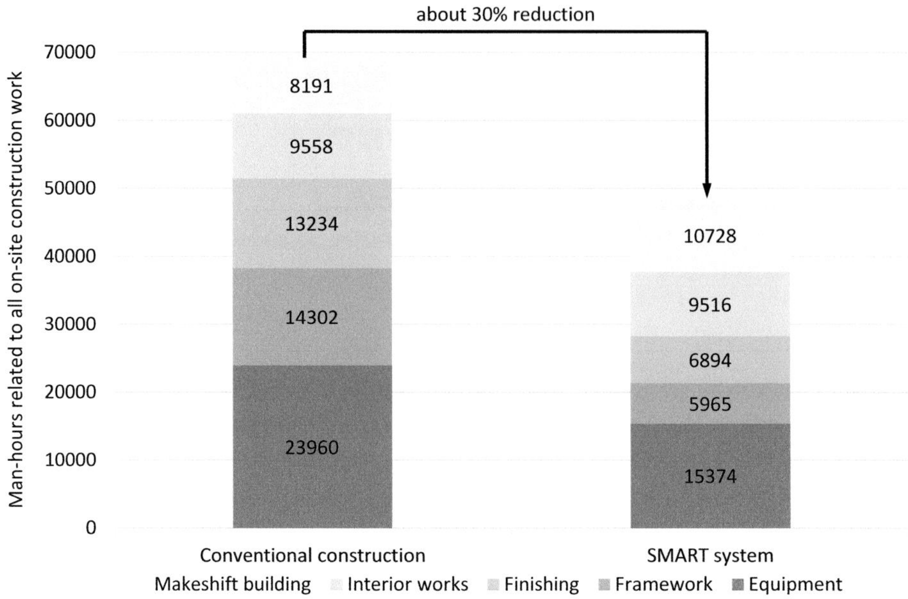

Figure 2.126. Reduction of man-hours related to the whole on-site construction work (adopted from documentary material received from Prof. Maeda/Shimizu; see also Maeda, 1994; Maeda & Miyatake, 1997).

Learning Effects Figure 2.127 shows that, during the first projects completed by SMART, the time for the installation (logistics from storage on ground plus positioning) of the bearing steel columns and beams was reduced from 42 minutes to 24 minutes (after about 20 cycles). The most considerable reduction took place during the first two cycles. Figure 2.127 also shows that (according to the reduction of installation time) the number of transport and positioning operations (installation operations) nearly doubled.

Reduction of Construction Waste SMART, in the first projects completed by it, reduced the amount of construction waste generated on-site by 70% as a result of

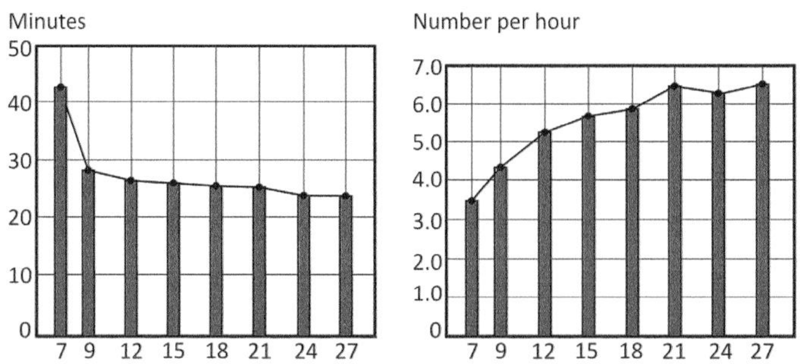

Figure 2.127. (Left) Improvement of times needed for the transportation/installation of structural components. (Right) Improvement of the number of transportation operations (adopted from documentary material received from Prof. Maeda/Shimizu; see also Maeda, 1994; Maeda & Miyatake, 1997).

the high prefabrication and the on-site material management system (Figure 2.128). A reduction of 70% is equivalent to a waste volume of about 700 tons. Construction waste still represents the biggest waste proportion compared to waste from manufacturing industries even in highly industrialized countries (see also **Volume 1**).

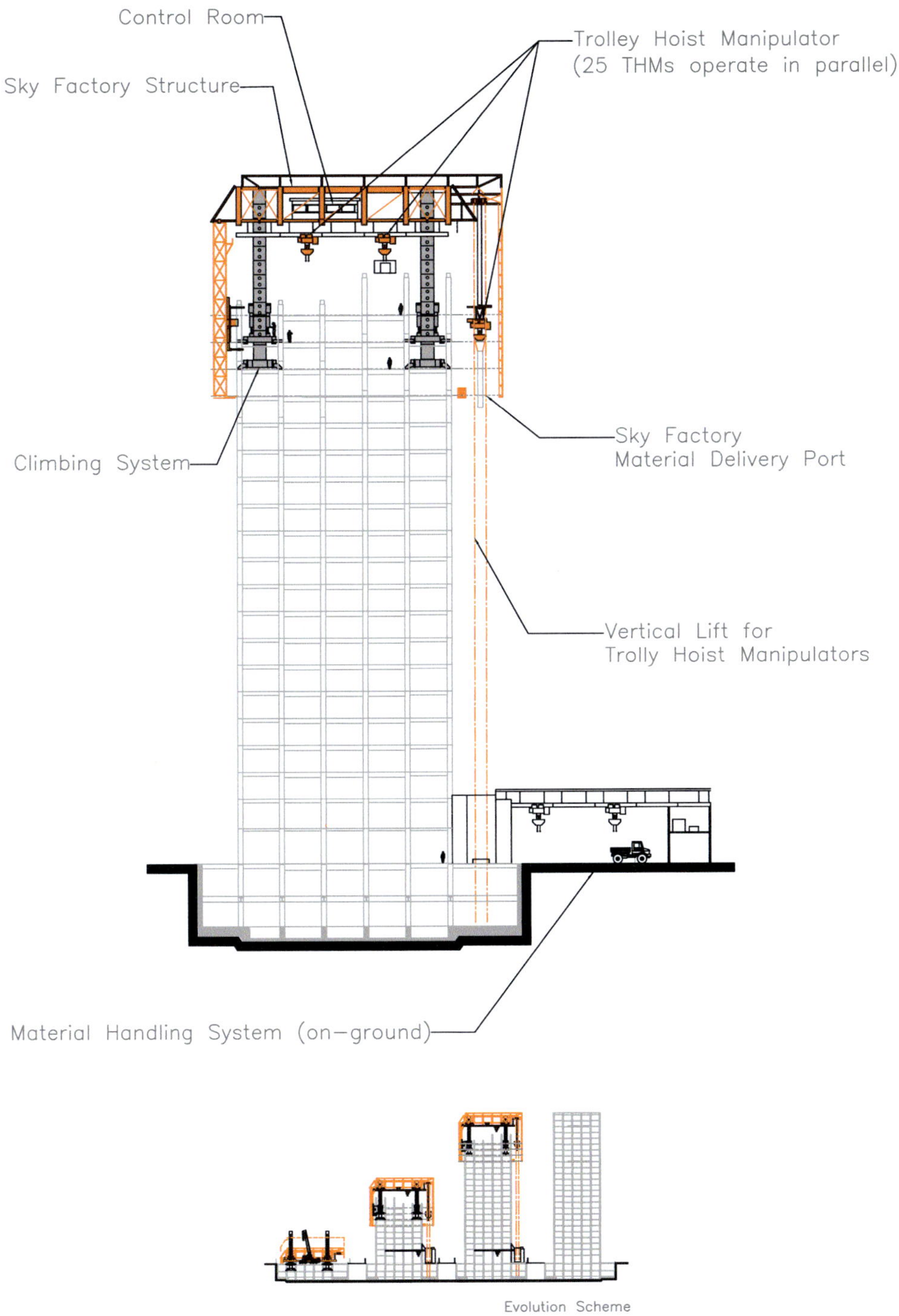

Shimizu Manufacturing System by Advanced Robot Technology (SMART), Company: Shimizu Corporation, Japan

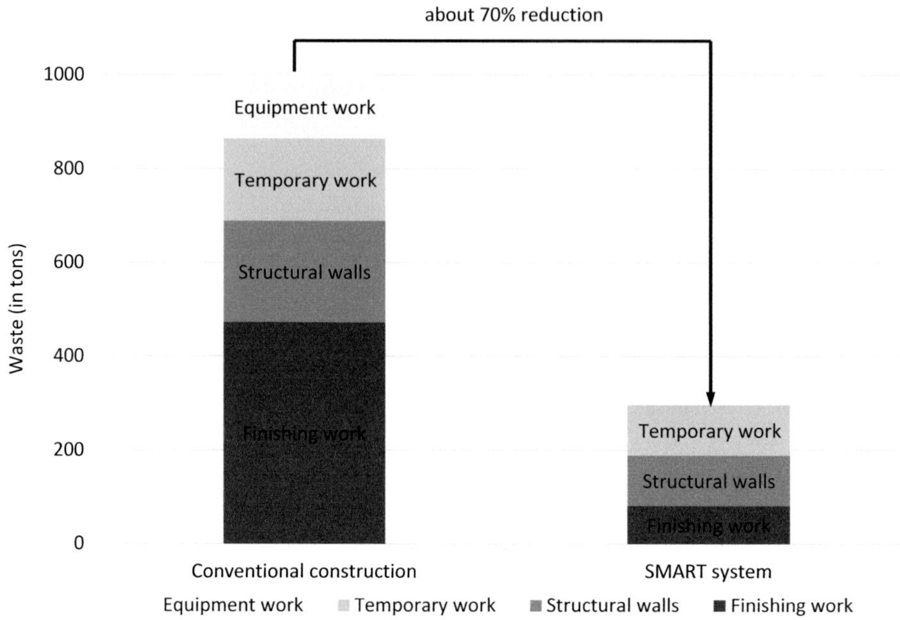

Figure 2.128. Reduction of construction waste realized through the application of SMART (adopted from documentary material received from Prof. Maeda/Shimizu; see also Maeda, 1994; Maeda & Miyatake, 1997).

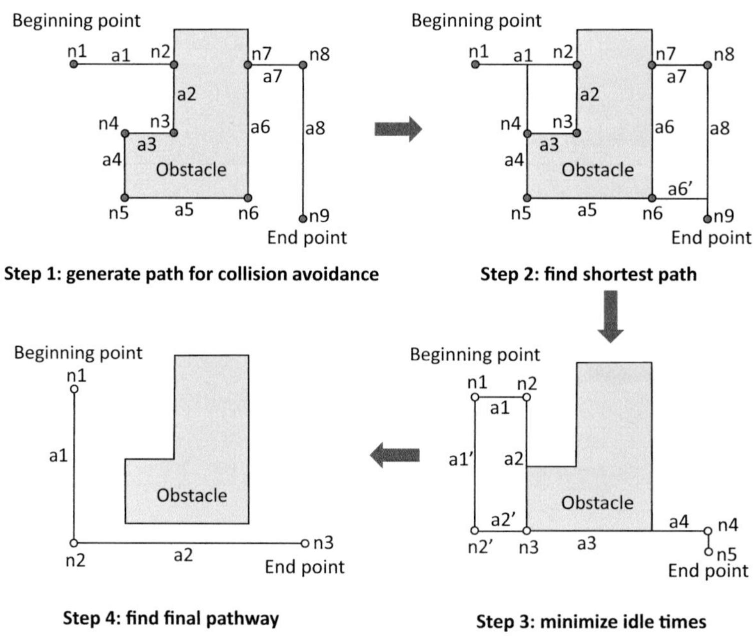

Figure 2.129. Generation of transportation paths and collision avoidance schemes (adopted from documentary material received from Prof. Maeda/Shimizu; see also Maeda, 1994; Maeda & Miyatake, 1997).

Product Quality and Process Monitoring Shimizu developed software for the SMART to generate and simulate transporting paths and positioning operations (Figure 2.129). Besides optimization of the transport and positioning operations, one aim of the system was to avoid collisions and thus damage to the building and building components. In addition to an assembly simulator, Shimizu developed a SMART-system disassembly simulator, which enables the simulation and optimization of the disassembly of the SMART on-site.

Safety The SMART on-site factory on top of the building covers the workspace completely and protects it from:

- Wind
- Rain
- Sun (and therefore overheating in summer)
- Cold weather in winter

The workspace's climate is modified by a ventilation system. Artificial light further ensures appropriate working conditions and offers the possibility that the site potentially can be operated day and night. Furthermore, as the smart factory completely covers the building, finishes floor-by-floor and provides safe, covered working platforms, even for facade work, dangerous situations (e.g., falls) are completely avoided, improving the concentration of the worker on the given work tasks, in addition to worker safety.

SMART Application Example 1

Figure 2.130. Sky factory in an advanced construction stage.

Figure 2.131. Sky factory in an advanced construction stage.

Figure 2.132. Sky factory in an early construction stage.

Figure 2.133. Sky factory encloses the part of the building in which construction is done completely.

Figure 2.134. Installation of a column component by an overhead manipulator.

Figure 2.135. Climbing mechanism actuating the climbing system's stilts.

Figure 2.136. Port in the sky factory for automated vertical component delivery from the ground factory.

SMART Application Example 2

Figure 2.137. Sky factory enclosing the upper end of the not-rectangular building.

Figure 2.138. Finished product.

Figure 2.139. Sky factory, exterior view.

Figure 2.140. Installation of the climbing system on the site prior to automated construction.

Figure 2.141. Installation of the sky factory on the site prior to automated construction.

Figure 2.142. Overhead manipulators.

Figure 2.143. Overhead manipulator transporting a column component.

Figure 2.144. Overhead manipulator transporting a column component.

Figure 2.145. Automated gantry crane rails provide a flexible mobility infrastructure for the overhead manipulators.

Figure 2.146. Overhead manipulators travelling along the rails of the gantry crane system.

Figure 2.147. Gantry crane system moving along rails installed on the ceiling of the sky factory.

Figure 2.148. Overhead manipulator travelling from one gantry crane rail to the other.

Figure 2.149. Overhead manipulator awaiting transport by an automatic lift from the ground factory to the sky factory.

Figure 2.150. Port through which overhead manipulators are transported into the sky factory.

Figure 2.151. View into the ground factory overhead manipulator awaiting transport into the sky factory.

Figure 2.152. Trucks deliver components just in time and just in sequence to the ground factory.

Figure 2.153. Multipurpose end-effector of the overhead manipulator.

SMART Application Example 3

Figure 2.154. Components of the SMART equipped with electronics and mechanics (e.g., gantry cranes and overhead manipulators) are stored inside a hall.

Figure 2.155. Components not equipped with electronics and mechanics (e.g., stilts) are stored outside.

Figure 2.156. Gantry cranes and rail system.

Figure 2.157. Gantry cranes and rail system.

Figure 2.158. End-effectors of the overhead manipulators.

2.1.8 System Netherlands

Company: Royal BAM Group, Netherlands
Category: SF (moving upwards) supported by building

Analysis of Technology and Composition

Evolution Scheme This system was introduced in the Netherlands by Royal BAM. The system can be considered as mechanized construction factory system with some elements of light automation in an SF (Figure 2.159). A special feature of the system is the high degree of prefabrication achieved regarding the components processed. Highly prefabricated wall and facade concrete-based components (for example, already containing windows, interior and facade finishing, etc.) are delivered to the site and installed by the OMs. After positioning and adjusting, these prefabricated components require only a small amount of additional work (e.g., sealing) so that the amount of work activities and thus the complexity in the SF is reduced and a structured work environment is created. The SF integrates two OMs with a capacity of 40 tons each.

Elevation The elevation reveals the interplay of site logistics, OMs, and climbing strategy in detail. The vertical transportation of components is performed by an OM. One end of the SF is extended beyond the building's outer wall and kept open for the VDS. One of the OMs picks the material from the ground logistics yard in front of of the building, transports it vertically into the SF, and then translates horizontally to place the material in the SF near the installation area. The second OM serves as HDS, is more dexterous, and has 4 DOFs and a revolving part that assists in positioning and aligning of the prefabricated components that were previously transported into the SF by the other OM/VDS.

Figure 2.159. Sky factory, exterior view.

Ground Plan Enclosed working platforms flank the building outside the outer wall of the building (but inside the SF's cover) and supported by the SF's structure. The system utilizes two OMs as its central element. Both are integrated into the SF structure. One OM is remotely controlled and has 3 DOFs and serves as on-site logistics system on each floor for material transportation from the intake port to the location/area where the component is assembled. Both OMs run on rails that are integral part of the SF's frame structure. One side of the SF is equipped with an intake port for vertical logistics by the VDS's OM.

Subsystems The System Netherlands manufacturing the building on-site by way of an SF consists of a set of integrated subsystems:

1. *Sky factory structure*. The SF (Figure 2.160) covers the work area on top of the building and creates an SE for both manual work activities and OM driven component assembly operations. Further complexity is reduced through the processing of high-level concrete components in the SF. The SF consists of a steel frame, which also integrates working platforms on various levels inside and outside (Figure 2.169) of the building. One side of the SF contains an intake port for material delivery.
2. *Climbing system*. It ensures that once the construction of a floor is completed the SF is raised and supported by truss-like structures on the basis of the already constructed floors.
3. *Vertical delivery system*. Building components are vertically lifted through an intake port into the SF by one of the OMs serving as VDS and system for rough positioning of the components near the installation location in the SF (Figure 2.163).
4. *Horizontal delivery system*. OM with 4 DOFs (Figures 2.161, 2.162, and 2.164). The OM (a modified conventional overhead crane) has a lifting capacity of 40 tons. It contains a translating and a rotating mechanism guided gantry-like between two rails for positioning and alignment support of the prefabricated high-level concrete components and runs on rails integrated into the SF structure.
5. *Real-time monitoring and management system*. BAM's system can be considered as a mechanized system with some automated elements.

End-effectors The OMs of both VDSs and HDSs are each equipped with fixed but different multipurpose end-effectors:

1. End-effector of vertical delivery system: Hoist with one multipurpose hook
2. End-effector of horizontal delivery system: Rotatable hoist with a set of hooks

System Variations Flexible *in situ* concrete placement in this system is combined with an extensive use of customized precast concrete elements with a very high degree of prefabrication (Figures 2.165–2.168). The low degree of automation on the site in combination with the relatively simple structure of the SF and the CS allow a customization of the system to almost any rectangular, vertically oriented, and grid-based building composed of concrete elements.

Robot-Oriented Design The wall components are prefabricated and the design and accurate off-site manufacturing of those components considers the efficient use of the HDS in the SF. The high degree of prefabrication also reduces the amount of work activities on the site and structures the environment in the SF.

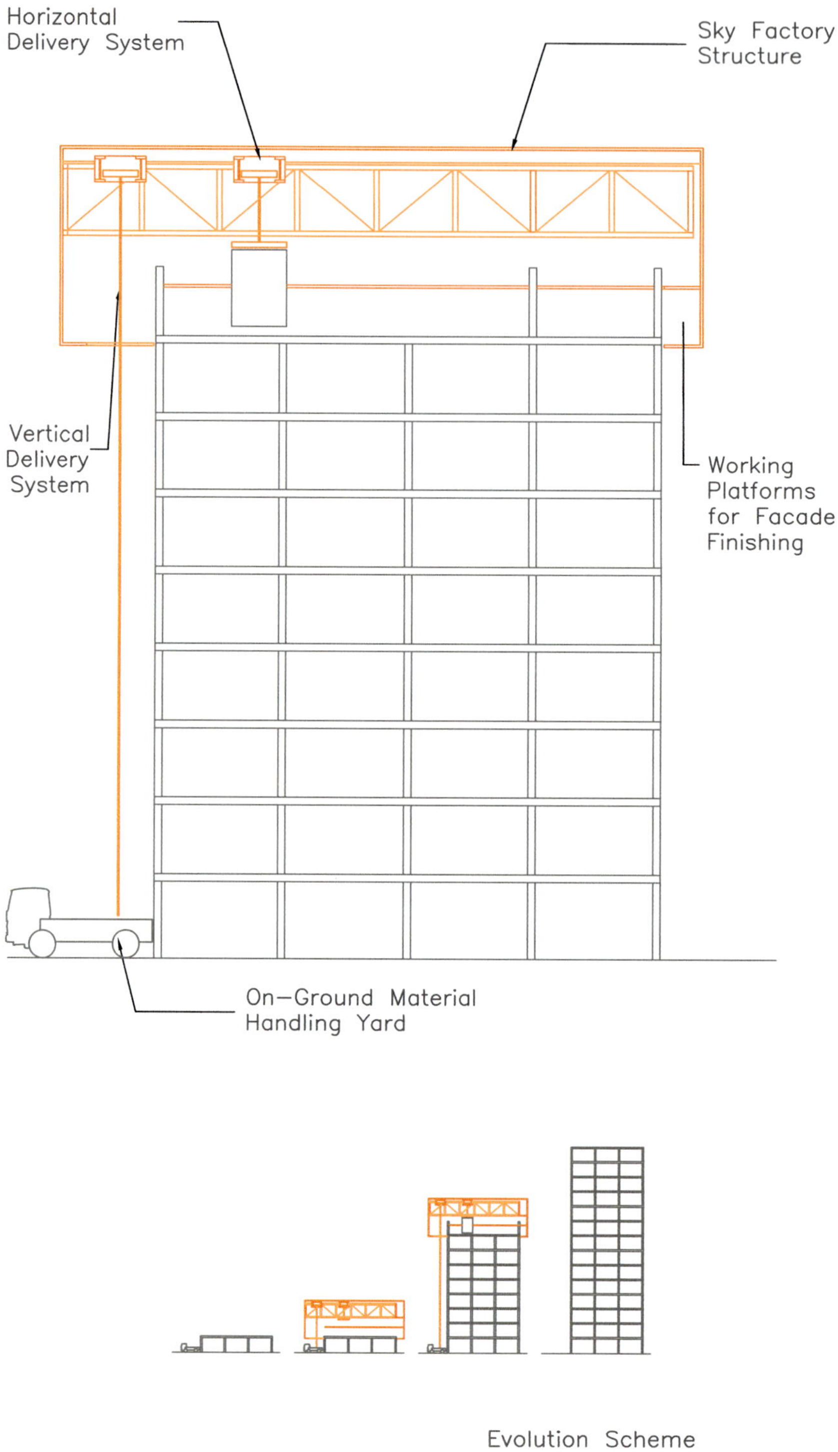

System Netherlands, Company: Royal BAM Group, Netherlands

Figure 2.160. Sky factory, exterior view.

Figure 2.161. Sky factory, interior view.

Figure 2.162. End-effector with multiple hooks.

Figure 2.163. End-effector with one hook.

Figure 2.164. End-effector with multiple hooks.

Figure 2.165. Use of mobile working platforms within the sky factory.

Figure 2.166. Intersection of two concrete components prepared for sealing.

Figure 2.167. Port for material delivery from the ground to the sky factory.

Figure 2.168. View from port to ground.

Figure 2.169. Working platform integrated into the sky factory allowing work on the exterior of the building.

2.2 Sky Factory (moving upwards) on Stilts (extending)

Like with the previous category, the systems described in this category are used to assemble buildings from bottom to top. The factory where component installation is done sits on top of the building. However, in contrast to the systems outlined in the former category, the SF does not rest on the structure of the building itself, but is supported by its own structure, which transmits the forces directly to the ground. Furthermore, in this category, SFs only provide a roof structure, not a completely sealed working environment. As shown in the efficiency analysis of the systems, this nevertheless considerably reduces the influence of sun, wind, and rain on the work environment. As SFs are not resting directly on the building and its structure or columns, the systems in this category are more independent from the building structure and require less integration with it which enhances the application scope and thus the utilization rate of the systems. Furthermore, through the simplification of the factory (roof only), the systems are less capital intensive.

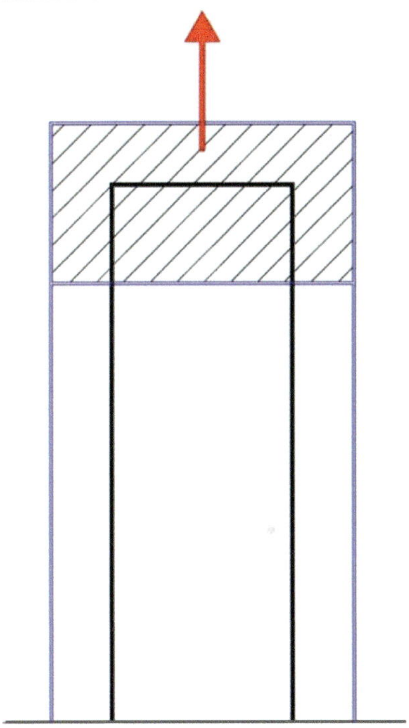

	System name	Company
9	Big Canopy	Obayashi
10	SMART Light	Shimizu

2.2.1 Big Canopy

Company: Obayashi Corporation, Japan
Category: SF (moving upwards) on Stilts (extending)

Figure 2.170. View into a sky factory.

Analysis of Technology and Composition

Evolution Scheme In the case of the Big Canopy (Figure 2.170), a roof structure is set up on ground level and then lifted over four steel stilts in its corners to an operational level. The system supports the construction of the first two levels and starts full operation from the third floor upwards. The stilts are located besides (i.e., not within) the building's structure and transfer loads generated by the SF and its subsystems to the ground. The stilts are composed of standard crane segments and are extended over a jib crane, which sits on top of the building. This feeds new stilts segments from top into the SFs roof structure so that the CS can lift the roof structure along the extended stilts. On disassembly of the SF, the roof structure is lowered in parts to the ground along the supporting stilts.

Elevation The elevation shows that the roof structure of the SF rests on a separate structure and, unlike systems of the former category, is constructed on top of the floors. This provides independence from the buildings column system and also allows the processing of heavy reinforced concrete elements. The SF is not completely covered but the rain and sun protection is considered to be sufficient. Rail-guided OMs are attached to the roof structure. An automatic lift (a variation of the automatic lift used in ABCS) accomplishes vertical logistics. The system is utilized to install medium level prefabricated concrete components (columns, beams, floor slabs, balcony components, facade components). Therefore, the buildings are modularized into concrete component kits. To attach components and seal gaps *in situ*, not prefabricated concrete was used in parallel.

Ground Plan The roof structure of the Big Canopy SF allows rail-guided OMs to be installed onto it. The main operation direction of the OMs is lengthwise and up to three OMs can be installed side by side. One automatic lift supplies the components to the OMs. The process of installing components (pickup, lifting, positioning, etc.) is simulated by a software tool and the operational sequence (and thus the configuration of the SF) is optimized by software tool.

Subsystems The Big Canopy system of manufacturing buildings on-site consists of many integrated subsystems:

1. *Sky factory factory roof structure*. The factory roof structure is made from a modular steel frame system and is adjusted for each new building (Figures 2.179, 2.180, and 2.192). The operational floors are protected by the roof structure but not completely covered. The control and structuring of the environment is thus unable to fully control all environmental factors, but provides more freedom, faster installation, and disassembly of the site factory and for a lower investment/fixed cost. As with other systems, rails guiding OMs are attached to the SF roof structure.
2. *Climbing system*. The factory rests and climbs; the CS is located in the roof structure and hydraulically pushes the roof structure upwards along four steel stilts (Figures 2.187 and 2.204). On disassembly of the SF these parts of the roof structure are lowered to the ground along the supporting stilts.
3. *Vertical delivery system*. An automatic lift (an adjustment of the automatic lift used in ABCS; Figures 2.188, 2.200, 2.203, and 2.205) accomplishes vertical logistics. Similar to the four steel stilts along which the SF climbs, the stilts along which the automatic lift runs can be extended by the jib crane on top of the factory or by the OMs.
4. *Horizontal delivery system*. The HDS is installed into the roof structure in the form of rail-guided OMs (Figures 2.181 and 2.186). Two completely different OM system configurations have been developed:
   a. Type 1: Two rail-guided manipulators with 5 DOFs (translation lengthwise, translation widthwise, rotation, forward/backward translation of main link, lower/lift of final link) and simple multipurpose end-effectors (Figures 2.199–2.207). The workspace of the en-effectors overlaps in the middle so that both manipulators have equally full access to the automatic lift.
   b. Type 2 (e.g., Big Canopy applied in Singapore and Chiba; Figures 2.183, 2.184, 2.185, and 2.189–2.198): Multiple hoist manipulators travel horizontally via an

overhead rail and gantry crane system. One high-speed rail-travelling delivery gantry crane sits in the middle. Two gantry cranes flanking the delivery gantry crane in the middle. The central delivery gantry crane can take up two hoist manipulators (one empty hoist coming in from a flanking gantry crane, having installed its component; one hoist carrying a component to be installed). With one operation the middle crane thus carries one manipulator hoist with a component to a flanking crane and in the same operation delivers an empty manipulator hoist over the automatic lift to pick up a new component. This results in a reduction of idle time or waste processes. As the travelling hoists themselves only allow the bringing of a component over the installation location for orientation robotic end-effectors (with an additional 2 DOFs; attached to the hoist manipulators), each gantry crane provides 1 DOF (translation lengthwise the building), and the hoist manipulator adds two more DOFs (translation, Lift/lower). So, in total, the HDS in this configuration provides 5 DOFs.

5. *Real-time monitoring & management system.* Software tools are used for the simulation and optimization of the operational sequence and the configuration of the system.

End-effectors The manipulators (HDS) can be equipped with various end-effectors:

1. HDS type 1: Simple multipurpose end-effector
2. HDS type 2: Robotic end-effector for component orientation (2 DOFs)

System Variations The independence of the system from the column grid structure makes it more flexible and easier to adjust (less preplanning and upfront investment needed). For experimental and optimization reasons, two types of HDS were developed: one with a manipulator approach and one with a travelling hoist manipulator approach.

Robot-Oriented Design The system has been utilized to install medium level prefabricated concrete components (columns, girders/beams, floor slabs, balcony components, facade components). Therefore, the building is modularized into a concrete component kit. To fix components and seal gaps *in situ*, concrete was used in parallel. The components are optimized in terms of their dimension, shape, and weight to suit the use of the OMs.

Efficiency Analysis

To reduce complexity on-site, Big Canopy processes prefabricated concrete components on-site. The building is thus divided into a kit of concrete components (walls, beams, columns, ceiling/floor elements, balcony elements). The rate of prefabrication accounted for about 70% of the concrete volume.

Detailed Project Schedule Big Canopy is able to erect concrete-based buildings. In the example of an erected building, given in Figure 2.171, the erection of the whole building including excavation and final finishing work took about 2 years (25 months). The Big Canopy system was applied from the 3rd floor to the 27th floor. It was able to construct 24 floors in 9 months, which is the equivalent of 2.7 floors per

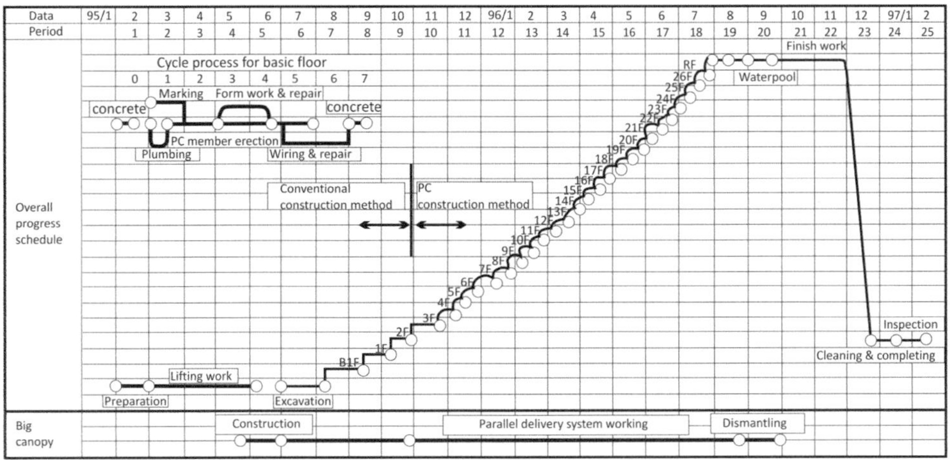

Figure 2.171. Overall progress schedule and FEC (adopted from Wakisaka et al., 1997).

month, on average. The set-up of the Big Canopy system took around 1 month, and the dismantling also 1 month.

Floor Erection Cycle With Big Canopy, in the preceding example, floors were erected on a 6.5-day cycle (Figure 2.171). It started with plumbing work on top of the previous floor, and then the prefabricated concrete elements were installed. In parallel, in areas where the concrete elements had already been placed on a floor formwork, wiring, and rebar work took place. At the end, a phase followed whereby on-site concrete was used to complete the floor.

Technical Data, Speed of Equipment In most applications of the system, the main contributors to overall system speed speed were an automated lift, the delivery OM, two erection OMs, and set of three travelling hoists. The travelling speed of the equipment equalled to 30–40 m per second (Table 2.5).

Experiments with Degree of Automation/System Configuration/Worker Teams Obayashi experimented with running Big Canopy in manual in automatic control modes (Figure 2.173). The analysis of the experiments shows that the manual operation is 5–10% faster. However, automatic control reduces the need for a supervisor and thus saves man-hours.

Productivity Big Canopy is able to enhance labour productivity (on average) by around 70% compared to conventional construction (Figure 2.174). The labour

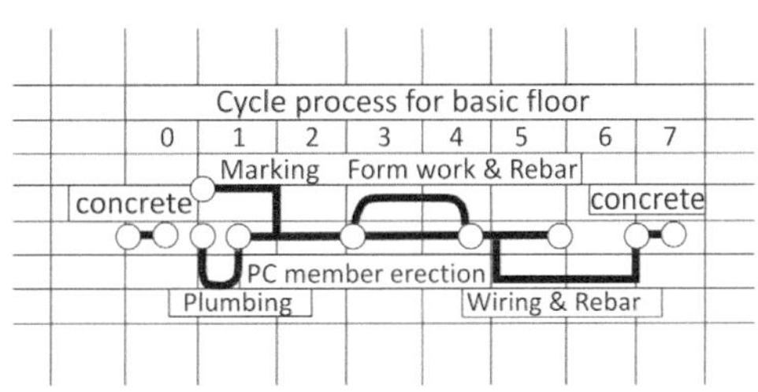

Figure 2.172. FEC (adopted from Wakisaka et al., 1997).

Table 2.5. *Specifications of parallel delivery system*

Name and number of device	Specifications
Construction lift/VDS	Loading capacity: 6 tons Winching up speed: 40 m/min Control type: inverter
OM used for hoist delivery	Operation type: manual/automatic wireless remote-control Control type: inverter
Delivery OM 1	Maximum travelling speed: 40 m/min
Erection OM 2	Suspension capacity: 7.5 tons
Erection hoist 3	Maximum travelling speed: 30 m/min Suspension capacity: 7.5 tons Maximum travelling speed: 33 m/min Suspension capacity: 7.5 tons
Gyroscopic suspender	Operation type: wireless remote-control Weight: 1100 kg Rotating drive: Gyroscopic moment Inertia moment of load: 25 tons/m

Adapted from Wakisaka et al. (1997).

productivity improvement, compared to conventional precast concrete construction, is around 10%, and compared to construction with formwork systems or kits it is 40%. Compared to conventional construction, Big Canopy particularly reduces the labour required in steel bar and formwork work. Compared to conventional precast concrete construction, Big Canopy significantly reduces the amount of work related

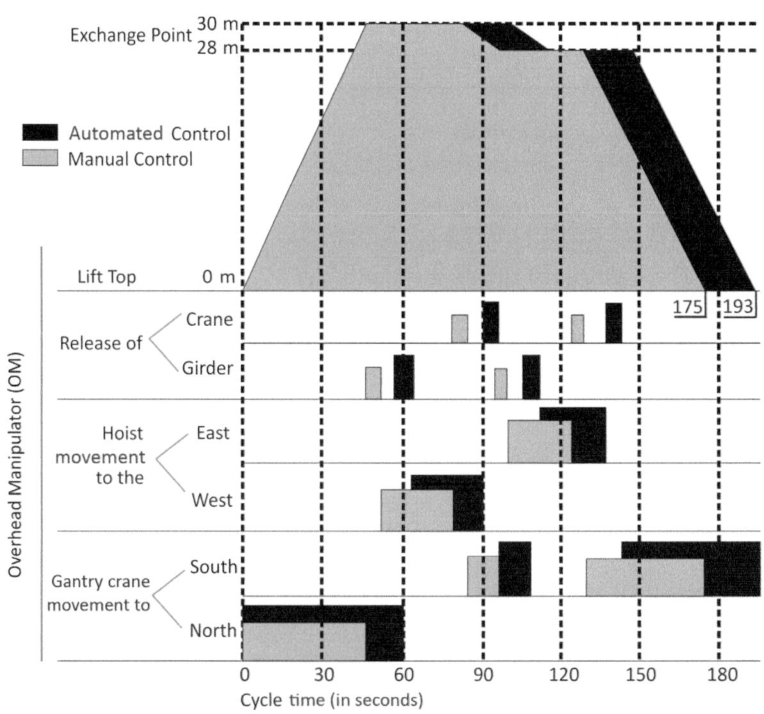

Figure 2.173. Comparison of the cycle times for various work activities between automatic and manual control modes (adopted from Wakisaka et al., 1997).

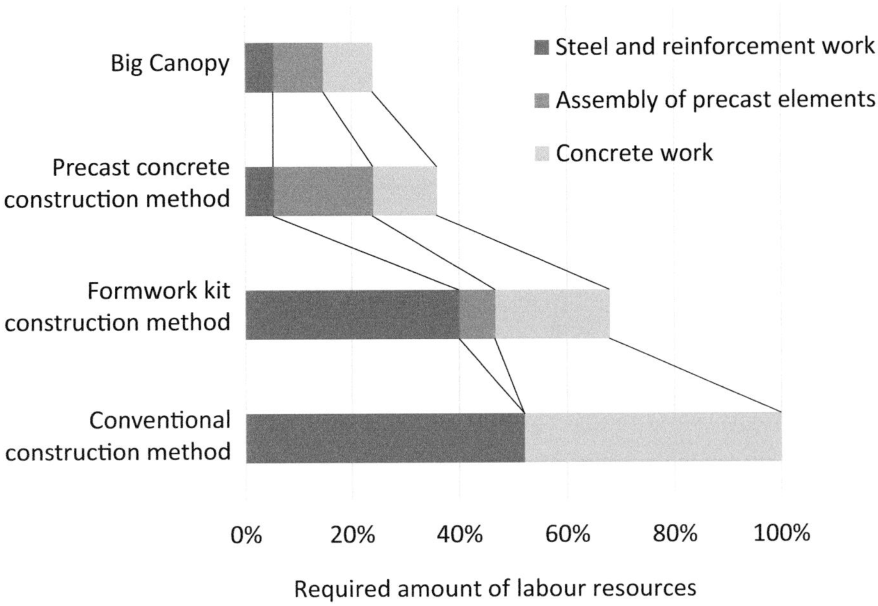

Figure 2.174. Comparison of work productivity for four construction methods (Wakisaka et al., 1997).

to the installation of the concrete elements. Furthermore, comparisons have shown that Big Canopy is faster at erecting a building than two tower cranes (Figure 2.175).

Learning Effects Analysis of the number of workers and man-hours needed to complete a standard floor within a Big Canopy SF shows that through learning effects, improvements to the main work tasks assisted by the system (floor slab installation, balcony element installation, column component installation, wall component installation and beam/girder installation) of up to 20% less workers/floor are possible. Considerable improvement starts from the tenth floor (Figure 2.176).

Product Quality and Process Monitoring To find an optimal sequence of logistics and installation operations conducted by Big Canopy's subsystems (e.g., VDS and HDS), a transport cycle-time simulator is used in advance of each construction

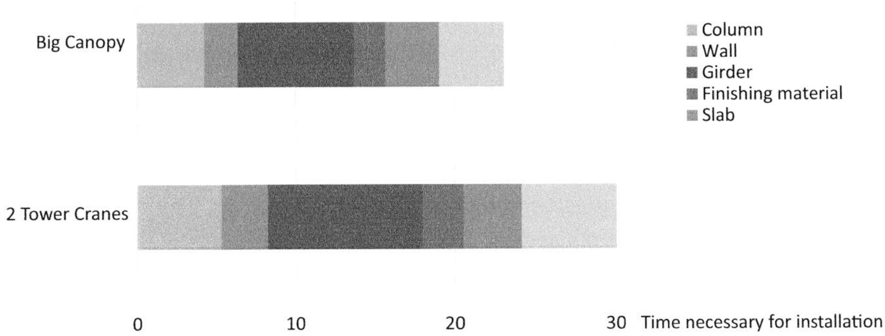

Figure 2.175. Efficiency comparison between a system with two tower cranes and Big Canopy system related to the installation of various types of components (adopted from Wakisaka et al., 1997).

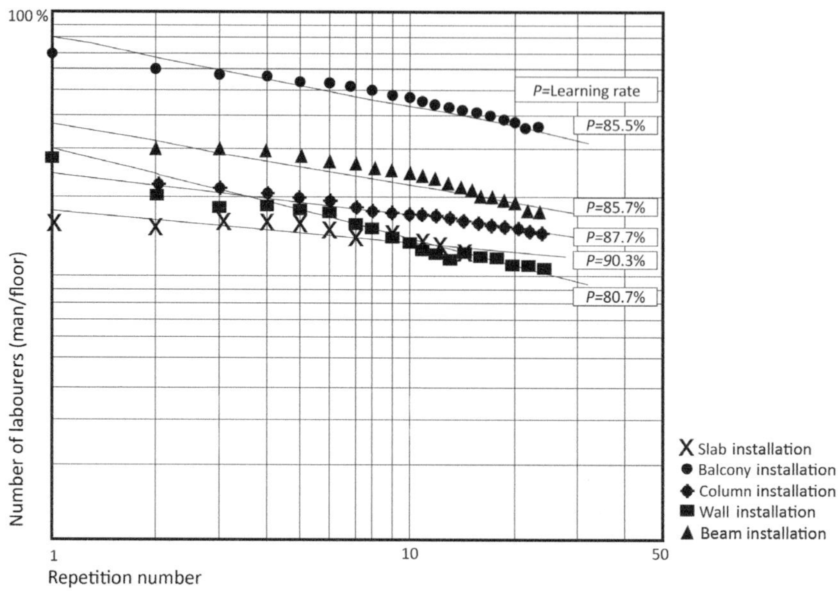

Figure 2.176. Reduction of amount of labour resources by learning effects (adopted from Wakisaka et al., 1997).

project. An RTMMS allows the partly automated supervising of material flow and work activities. The RTMMS consists of the following subsystems:

- Material Plan Subsystem
- Material Management System
- Actual Management System
- Erection Site Management System

The RTMMS has similarities to, for example, Enterprise Resource Planning (ERP) systems common in the manufacturing industry. It allows the setting up, optimization, and monitoring of material flows and work activities as well as the control of the progress of the overall manufacturing process. As all materials are tagged (barcoded), the flow of material, as well as all assembly operations executed by Big Canopy and thus work activities become transparent what allows for a real-time control of the construction progress and construction quality achievement.

Safety The modification of the working environment enhances both safety and productivity. Obayashi conducted studies on how the temperature influences productivity on-site. Japan is a country where, especially in summer, it is hot and humid. Homes, offices, and factories are usually equipped with climate regulation systems to mitigate the influence of the temperature. By erecting a temporary roof over the site, Big Canopy also tries to mitigate the influence of temperature.

Figure 2.177 analysed a standard building constructed by Big Canopy and shows that both the workers and the materials (such as reinforcement) were protected from overheating through exposure to sunlight. Obayashi further calculated that during the measurement period from July to September (123 days), the temperature reached greater than 30°C on 66 days at conventional sites with no protecting roof and accounted for about 50 hours (equivalent to 6.3 working days) of lost working

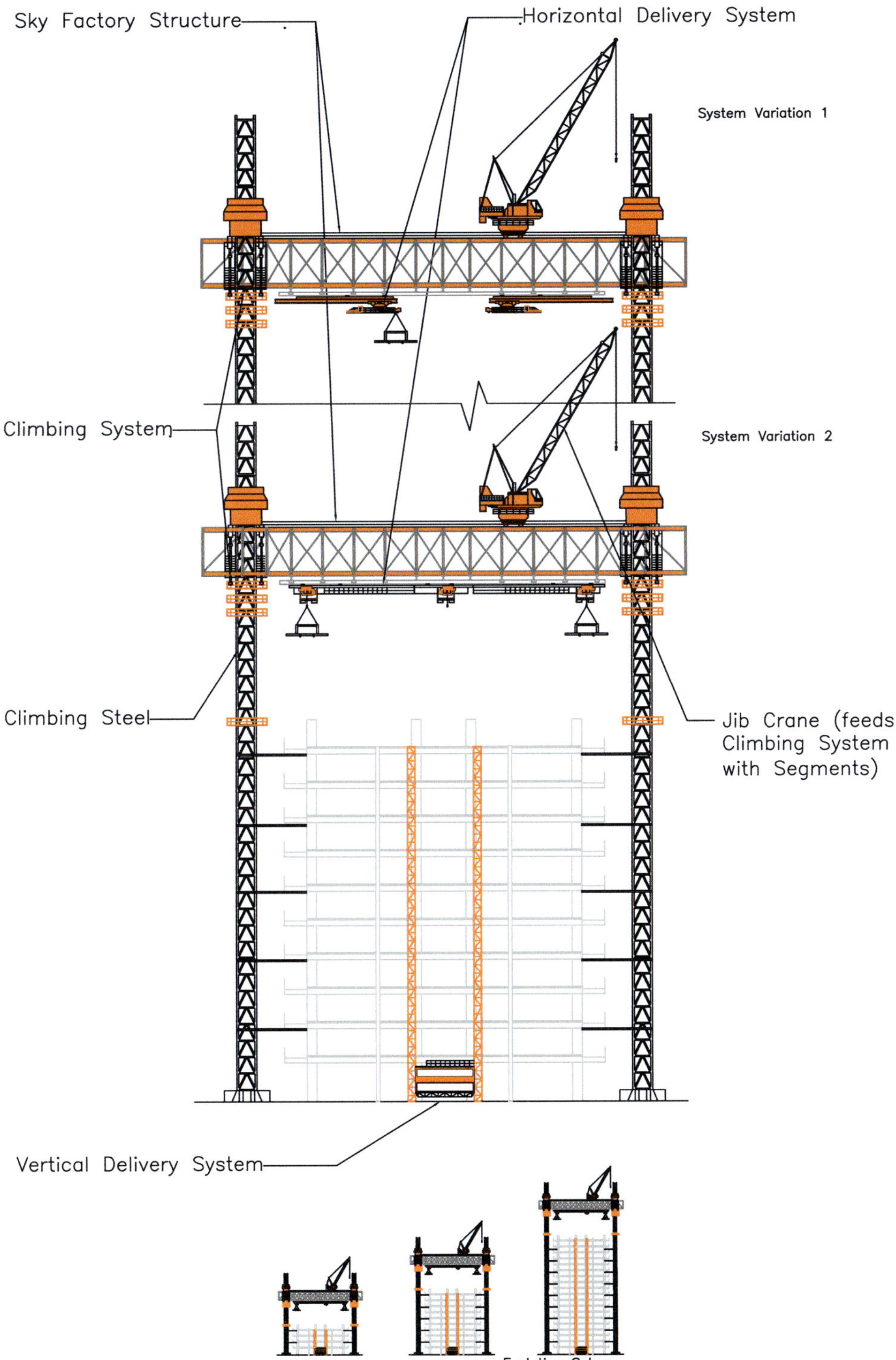

Big Canopy, Company: Obayashi Corporation, Japan

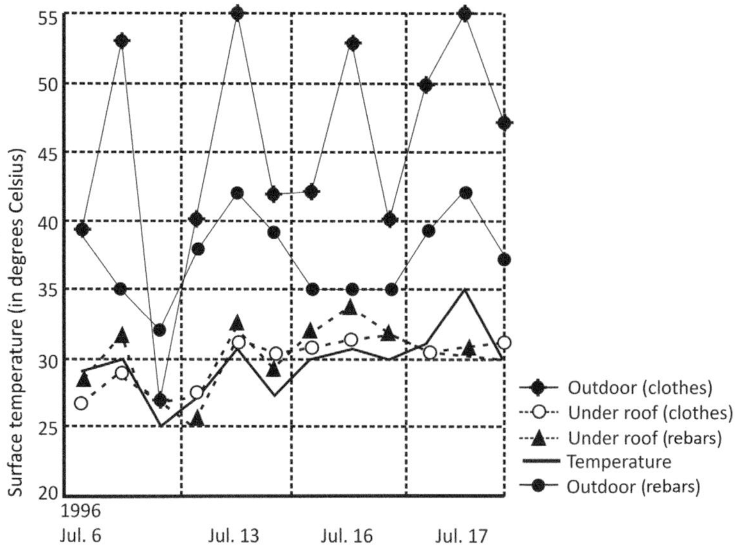

Figure 2.177. Comparison of surface temperature of workers' clothes and reinforcement between SF (under roof) and (outdoor) conventional construction (adopted from Wakisaka et al., 1997).

time, equivalent to an overall decrease of productivity of 10%. Big Canopy, and other systems with a protective roof, can reduce this type of productivity loss to practically zero. Furthermore, the reduced temperature also decreases both the mental and physical strain workers are subject to, enhancing concentration and thus safety and quality.

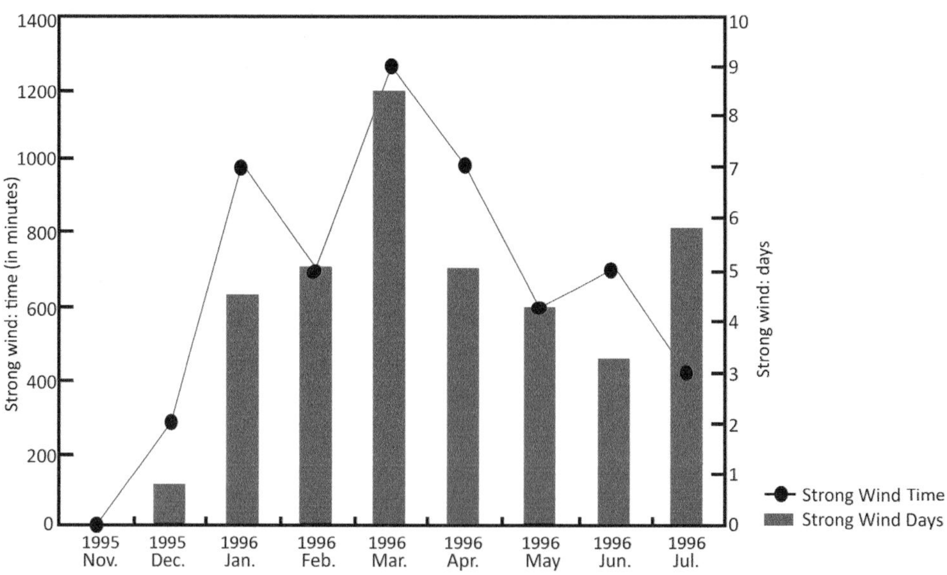

Figure 2.178. Wind condition measurement and determination of strong wind hours (Japan, urban area) and days that could influence and slow down conventional construction (adopted from Wakisaka et al., 1997).

Figure 2.179. Installation of sky factory and climbing system.

Physical Strain To analyse the physical strain workers are subject to in conventional construction environments and in the Big Canopy working environment, Obayashi measured the heart rates of workers.

Heart rate figures of workers on conventional noncovered construction environments:

- Maximum: 134 beats/minute
- Mean value: 103 beats/ minute

Heart rate figures of workers in the Big Canopy working environment:

- Maximum: 108 beats/minute
- Mean value: 89 beats/ minute

According to Obayashi, in conventional construction the heart rate was 80%/40% higher than that measured during breaks. On the Big Canopy site it was only 20% to 40% higher.

Influence of the Weather Besides the above mentioned influence of the sun and temperature on the construction process, Obayashi analysed the influence of wind on the construction process (Figure 2.178). Obayashi continuously measured the wind's strength and impact on the building and SF (measured in minutes) during Big Canopy's application (precast concrete element positioning for standard floors). Obayashi found through this analysis that during Big Canopy's application, the wind speed exceeded about 10 m/s for 87 working hours, which equals to about 12 working days. A wind speed of about 10 m/s usually means that conventional tower cranes cannot be operated (due to so called "strong wind hours"). Big Canopy's SF reduced the wind speed related downtime significantly to 1.5 days.

Big Canopy Application Example 1

Figure 2.180. Sky factory overspanning the building.

Figure 2.181. Female construction worker operating an overhead manipulator.

Figure 2.182. Stilt bearing the sky factory.

Figure 2.183. Horizontal delivery system.

Figure 2.184. Horizontal delivery system.

Figure 2.185. Horizontal delivery system.

Figure 2.186. Sky factory, exterior view.

Figure 2.187. Climbing mechanisms.

Figure 2.188. Vertical delivery system.

Figure 2.189. High-level prefabricated components sequenced on the construction site.

Figure 2.190. Concrete components after assembly by overhead manipulators.

Figure 2.191. High-level prefabricated components sequenced on the construction site.

Big Canopy Application Example 2

Figure 2.192. View into the sky factory, early construction stage.

Figure 2.193. Overhead manipulator transporting a column component.

Figure 2.194. View into the sky factory, late construction stage.

Figure 2.195. Overhead manipulators switching between automated gantry cranes.

Figure 2.196. View on floor assembled by overhead manipulators.

Figure 2.197. Concrete components are either directly cast on site or delivered just in time and just in sequence.

Figure 2.198. Truck delivering concrete components just in time and just in sequence.

Figure 2.199. Overhead manipulator.

Figure 2.200. Vertical delivery system delivers a component to the overhead manipulator.

Figure 2.201. View into a sky factory.

Figure 2.202. Overhead manipulator manipulating a component.

Figure 2.203. Vertical delivery system, side view.

Figure 2.204. Climbing mechanism.

Figure 2.205. Vertical delivery system, front view.

Figure 2.206. Concrete components after assembly.

Figure 2.207. Concrete components after assembly.

Figure 2.208. Component awaiting transport by the vertical delivery system.

2.2.2 SMART Light

Company: Shimizu Corporation, Japan
Category: SF (moving upwards) on Stilts (extending)

Analysis of Technology and Composition

Evolution Scheme The SMART Light (Figures 2.209 and 2.210) is an altered and simpler version of the SMART. The SMART Light does not cover the workplace with a roof or side sheets but embraces the building with a steel frame, allowing various delivery systems and manipulators to be mounted to it. The steel frame, as well as the steel stilts on which the frame is lifted, is composed of standardized frame segments. The steel stilts on which the frames are mounted are an independent building structure erected besides the building. They transfer the load to the ground, making the erection of the steel frame independent from the building's column grid. The system can be erected on the ground floor over the floor area and can support the erection of the lower floors or even basement floors.

Elevation OMs are installed on the steel frame. Various jib cranes can also be mounted on top of the steel frame. The vertical delivery of parts and components is accomplished via the jib cranes. The vertical stilts stand right beside the building, and to achieve stability are connected to the building by cross beams approximately every fourth floor.

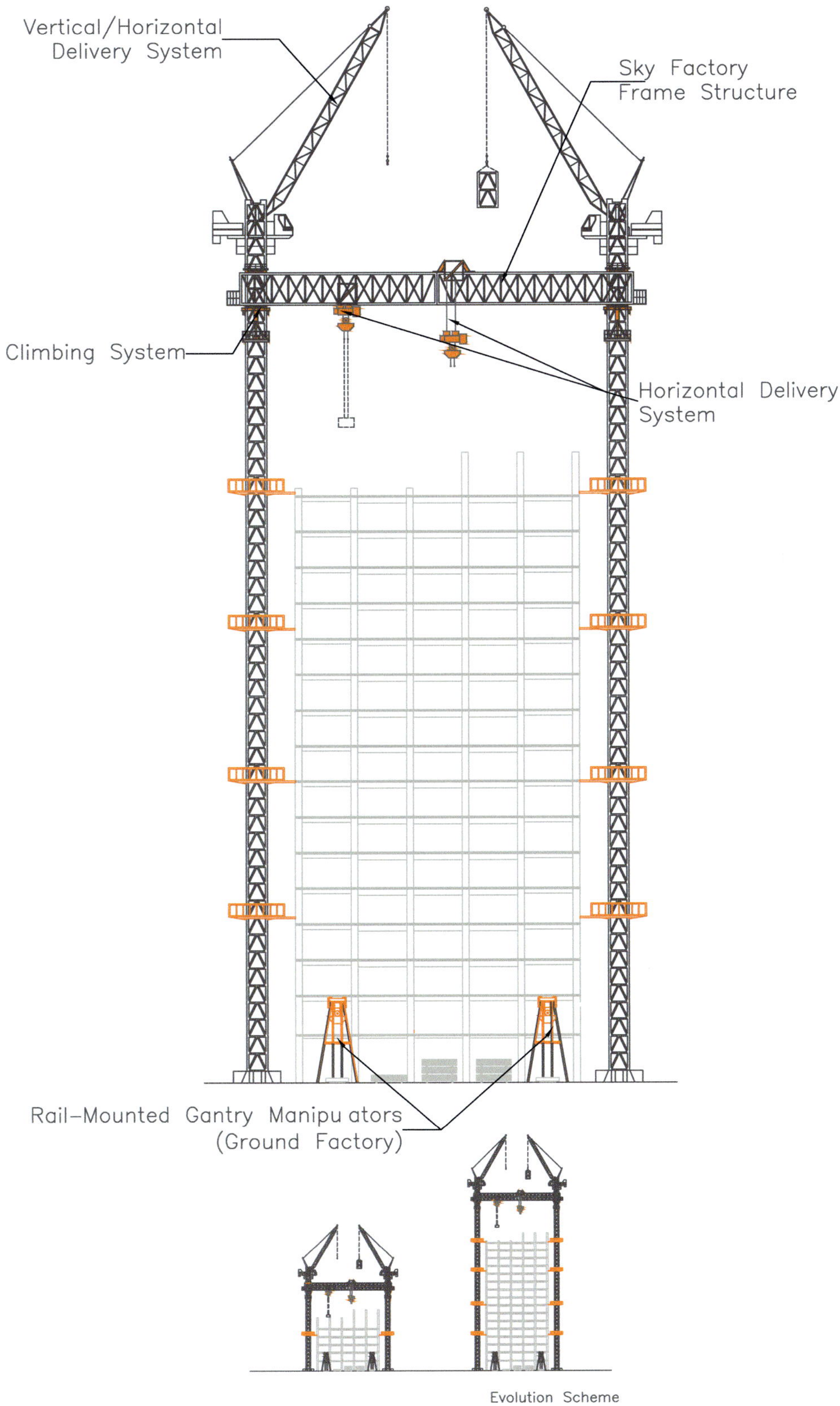

SMART Light, Company: Shimizu Corporation, Japan

Figure 2.209. View into a Sky factory.

Ground Plan The ground plan shows that the steel frame spans the building in a way that allows one to three gantry-type OMs to be installed, so that the workspace of all manipulators together covers the buildings floor area. The jib cranes on top are installed on top of the frame and in the middle of each of the shorter sides of the building. The workspace of the jib cranes spans the whole building and allows the lifting up of parts and components from delivery or storage locations on the ground. As the frame structure is not covered by a roof, the jib cranes and OMs can operate in parallel.

Subsystems The SMART Light on-site building manufacturing system consists of many integrated subsystems:

1. *Frame structure*. The frame structure consists of standardized segments and provides a platform for the installation of various OMs and jib cranes (Figures 2.211, 2.212, 2.219, and 2.220).
2. *Climbing system*. The CSs (Figures 2.214–2.216) consist of a hydraulic actuated inchworm-like CS, which raises the frame structure to which manipulators and cranes are mounted along the stilts of. To lengthen the stilts, the jib cranes place a new segment on top of the end of each of the stilts.
3. *Vertical/horizontal delivery system*. Jib cranes mounted to the steel frame operate in parallel to the OMs and accomplish vertical as well as horizontal delivery tasks.
4. *Horizontal delivery system*. Gantry-type OMs with 3 DOFs (translation lengthwise, translation widthwise, lower/lift of end link; Figures 2.213 and 2.222). For reasons of compactness, and to avoid collisions, two outer manipulators are guided by rails at the bottom of the frame, the middle manipulator by rails on top of the frame.
5. *Movable roof cover segments (optional)*. Movable roof cover segments (Figures 2.217 and 2.218) can be mounted to the top of the frame, covering some dedicated work places temporarily, as well as parts of the workspace of the OMs. For delivery of components by the jib cranes, the roof formed by this cover segments can open automatically.
6. *Real-time monitoring and management system*. The control stations/rooms for the manipulators, as well as for the jib cranes, sit on top of the frame. The gantry-type OMs can be operated in automatic mode as well as via remote control.
7. *Ground factory parts/components MHSPY*. Small sized and self-supported gantry cranes provide support with material handling, sorting, and logistics preparation on the ground (Figure 2.221).

End-effectors Both the overhead gantry-type manipulators and the vertical HDS use simple (conventional crane-like) multipurpose end-effectors.

System Variations As both the steel stilts and the horizontal frame are composed of standardized linear frame elements, and as the system is independent from the column grid structure of the building. It is highly flexible. Widthwise, one to three gantry-type manipulators can be installed. Lengthwise, the length of the frame girders can be varied. Furthermore, by adding two new stilts, a second frame with an additional one to three manipulators can be added to the lengthwise direction, so that "long" and horizontally oriented buildings can also be erected.

ROD An advantage of the system is its independence from the building grid structure, its flexibility/variability and its simplicity, which allows fast set-up. Owing to its independence, as well as the multipurpose nature of the end-effectors, the necessity to redesign components is almost removed. However, the system cannot achieve the same operational speeds or automatic capability of the SMART.

Figure 2.210. Variation of the system overspanning a large building.

Figure 2.211. Stilts bearing the sky factory.

Figure 2.212. Sky factory, side view.

Figure 2.213. Gantry-type overhead manipulators.

Figure 2.214. Climbing mechanisms.

Figure 2.215. Climbing mechanisms.

Figure 2.216. Climbing mechanisms, detailed view on actuator.

Figure 2.217. Movable roof cover segments.

Figure 2.218. Movable roof cover segments.

Figure 2.219. View into a sky factory.

Figure 2.220. View into a sky factory.

Figure 2.221. On-ground material handling yard.

Figure 2.222. Gantry-type overhead manipulator.

2.3 Sky Factory Pulled Up by Core (main factory and core factory moving upwards)

Systems in this category of buildings are assembled from bottom to top. The lifting of the SF is accomplished by pulling it upwards along a central core structure (steel or concrete), which is assembled a few floors in advance. In the case of Taisei's Totally Mechanized Construction System for High-rise Buildings (T-Up system; for more detail see the following section), three on-site factories are deployed thus:

1. A GF, which converts parts and components into medium or higher level components
2. A core SF around the building's core, which manufactures the core structure in advance and which sits on top of the main SF
3. The main SF, which is pulled upwards along the core structure

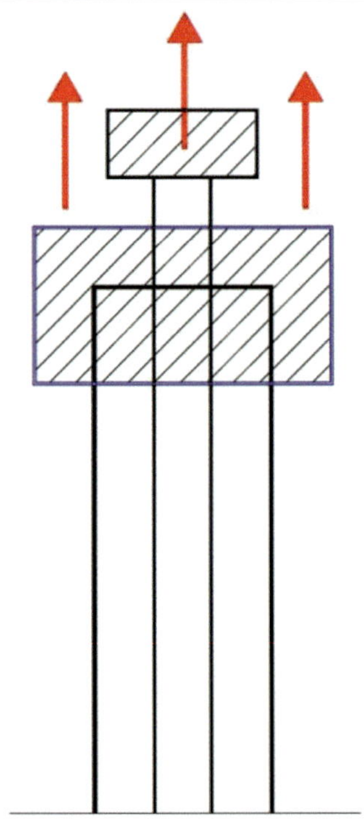

	System name	Company
11	Totally Mechanized Construction System for High-rise Buildings (T-Up)	Taisei
12	Robotic and Crane-based Automatic Construction System (RCACS)	Korean Consortium

Systems in this category are limited to the assembly of buildings with a central (structural) core. They utilize the characteristics of this building typology for the purpose of automation. As the core itself is often a simple and highly repetitive structure, its assembly can easily be systemized. The manufacturing of the building's core by the core SF fulfils the characteristics of a "line construction site" (for more information on the mechanization and automation of line construction sites by TBMs; see **Volume 1**) and is characterized by mainly vertical logistics and installation activities along the core. In contrast, the larger main factory (which is pulled upwards along the core produced by the core SF) has higher variability of functions as well as a higher integration of vertical and horizontal logistics and installation operations.

2.3.1 Totally Mechanized Construction System for High-Rise Buildings (T-Up)

Company: Taisei Corporation, Japan
Category: SF pulled up by Core (main factory and core factory moving upwards)

Figure 2.223. Sky factory, exterior view.

Analysis of Technology and Composition

Evolution Scheme In the case of the T-up system (Figure 2.223), a part of the SF is dedicated to manufacturing the core of the building. It does this automatically and then provides a basis for the automatic raising of the main SF in which the automatic manipulators operate. The T-up system thus visualizes the principle of robotic self-reproduction. The relevant part of the structure for the SF is the uppermost floor of the building, which is erected and finished on the ground and then pushed up along the core structure.

Elevation The elevation shows the interplay between the smaller SF which houses the climbing mechanism (hereafter SF part 1), and the main SF (hereafter SF part 2), which sits below the smaller of the two. In the following overview of the system, the term "SF" is used to refer to the two parts combined.

If a full GF is installed on the ground, the whole process of transforming low-level components into higher level components is fully automated on-site, as is the delivery to and installation of the transformed components in the SF.

Ground Plan The ground plan shows the working area of SF part 1 and SF part 2 robotic manipulators. In SF part 2, robotic trolley manipulators travel along an integrated rail-guided, gantry crane system. The trolley manipulators integrate vertical and horizontal delivery, thus erasing the need for the handover of components between vertical and HDSs. Component-type adjusted end-effectors simplify pickup and installation processes.

Subsystems The T-Up system manufacturing the building on-site consists of many integrated subsystems:

1. *Sky factory structure*. Covers the workplace and creates an SE. In the case of the T-Up system, two SFs are arranged one above the other (Figures 2.228–2.232).
   - *Factory part 1 (upper/core SF)*. Factory part 1 provides an SE for the manufacturing of the building's core, which enables the climbing of the whole SF. The factory is equipped with side covers and integrated with rail-guided jib cranes as part of the OM system. It is the platform for the installation of the measurement and steel column alignment system.
   - *Factory part 2 (lower SF)*. Factory part 2 provides an SE for the erection of the building around the building's core and the installation of facade elements. Unlike the factory part 1, factory part 2 is not equipped with side covers. It is the platform for the installation of OMs. The relation between parts 1 and 2 is symbiotic; part 1 manufactures the core of the building in advance to enable the operation and upward climb of factory part 2 (and with it part 1 also). The SF platform, which forms the basis for factory part 1 and 2, is also the uppermost floor of the building, and after completion is integrated into the building itself.
2. *Climbing system*. The CS lets SF part 2 climb up the inner core manufactured within factory part 1. Hydraulically actuated climbing stilts climb inchworm-like along every second steel column of the inner core, and therewith carry up SF (both parts). In total, taking into account all climbing stilts, the CS has a lifting capacity of 2000 tons.

3. *Vertical/horizontal delivery system 1 (jib crane manipulators).* On top of the SF's main platform (part of factory part 1), two rail-guided robotic jib cranes are installed. These lift and position the steel parts and components.
4. *Vertical/horizontal delivery system 2 (OMs).* The main element of factory part 2 is a system of rail-guided OMs (Figures 2.232, 2.234, 2.235, and 2.237). Two hoist manipulators, each capable of lifting parts and components of up to 10 tons, are able to travel along rail-guided automatic overhead cranes to any position within factory part 2. The gantry crane system allows the manipulators to travel to cantilever positions along both the sides and ends of the factory, and thus to directly pick up parts and components from the ground or GF. In combination with the automatic overhead crane system, each manipulator has 3 DOFs. Various component specific end-effectors can be attached to the manipulators, adding additional DOFs.
5. *Interior finishing systems (optional).* Taisei can use its STCRs (logistics and interior finishing robots) in finished and weatherproofed floors under the SF to finish the interior of the building.
6. *On ground horizontal delivery system (ground factory, optional).* On the ground, Taisei can optionally install a rail-guided AGV for the transport of material from trucks to the on-site building component manufacturing (BCM) line, to storage or to the component pick-up tables.
7. *On-site building component manufacturing line (GF, optional).* Taisei has developed a modular and temporarily installable manufacturing line for the concrete-based floor and facade elements. The manufacturing line can deliver components directly to the pick-up tables (Figures 2.233 and 2.236).
8. *Component pickup table (GF, optional).* Along the side of the building, rail-guided automated component pickup tables position components for collection by the V/HDS 2. The component pick-up tables orientate the components to be picked up in a standardized way and thus simplify an automated pick-up by the OMs. They are a kind of factory internal component palletizing system that simplifies FIL.
9. *Underground excavation system (optional).* Taisei can operate automatic excavation and soil removing systems in parallel to the operation of the SF to facilitate the construction of multiple basement levels.
10. *Alignment and accuracy measurement system.* A laser-based, automated system for measuring and aligning steel columns is installed in SF part 1. A precisely manufactured inner core reduces alignment complexity in the lower of the two factories (part 2) and guarantees precision in all following manufacturing sequences.
11. *Real-time monitoring and management system.* The on-site material flow (e.g., between GF and SF) and the progress of the installation of components is supervised in a control room located on the SF's main platform structure. In addition, on the basis of computer-aided design (CAD) data and feedback from the operation of the equipment, the progress of construction on-site can be monitored. The operation of the OMs and the component installation is graphically visualized and monitored in real time. Workers are equipped with handheld devices (smartphones, tablet computers) where they can view and thereafter accept all

necessary work tasks. The handheld devices are equipped with barcode or radio frequency identification tag (RFID) readers to allow workers to identify and track components.

End-effectors The manipulators (V/HDS 2) can be equipped with various end-effectors to enable the picking up, positioning, and adjustment of various component types:

1. End-effector for picking up, positioning, and adjustment of columns
2. End-effector for the positioning of beam/girder elements
3. End-effector for picking up and positioning concrete floor slabs
4. End-effector for positioning of prefabricated large-scale facade elements

System Variations The T-Up system can be applied in different configurations. SF parts 1 and 2 can be connected (for the erection of buildings with a steel-based inner core) via a raised roof platform, on top of which robotic jib cranes carrying robotic OMs would be mounted. If a building with a concrete core is erected, factory parts 1 and 2 are disconnected. Part 1 would then completely enclose the automatic concrete core production by automatic slip forming technology. The cover of factory part 1 can also be equipped with an OM, as an option. Another option is the installation of a GF to automatically finish and provide material to the SF.

Robot-Oriented Design On a component level, ROD is used to synchronize the dimensions, shape, and weight of the components to be delivered to and processed in the SF. To reduce complexity in the SF, (medium level) components can be prefabricated off–site or in an optionally installable GF, as mentioned previously. Pickup tables standardize the collection of components and further allow the complexity of the installation process to be reduced. On a building level, it can be decided in the design phase to apply the system only to buildings with a core structure; at this stage it can ideally be ensured that the building will have a near quadratic ground plan in which components will be organized orthogonally. On an urban level, it should be ensured in the design phase that the area around the building will be free-standing, and that space will be available around the building for pickup tables.

Efficiency Analysis

During one of the first projects completed by the use of the T-Up, Taisei conducted a detailed analysis of the performance of the system (Figure 2.224). The set-up of the core production structure takes about 1 month and the set-up of the structure for producing the core's periphery another 1.5 months. However, parallel to the set-up of the core's periphery production system, the core production structure can already be operated and can start to produce the first floors. The production of standard floors by the core's periphery production system starts after the second floor. The system is able to produce 28 floors in 6.5 months (without utilities and finishing work). The finishing of the floors takes another 4 months, and thus the completion of all 30 floors above ground floors takes 12.5 months in total, equal to 2.4 floors/month.

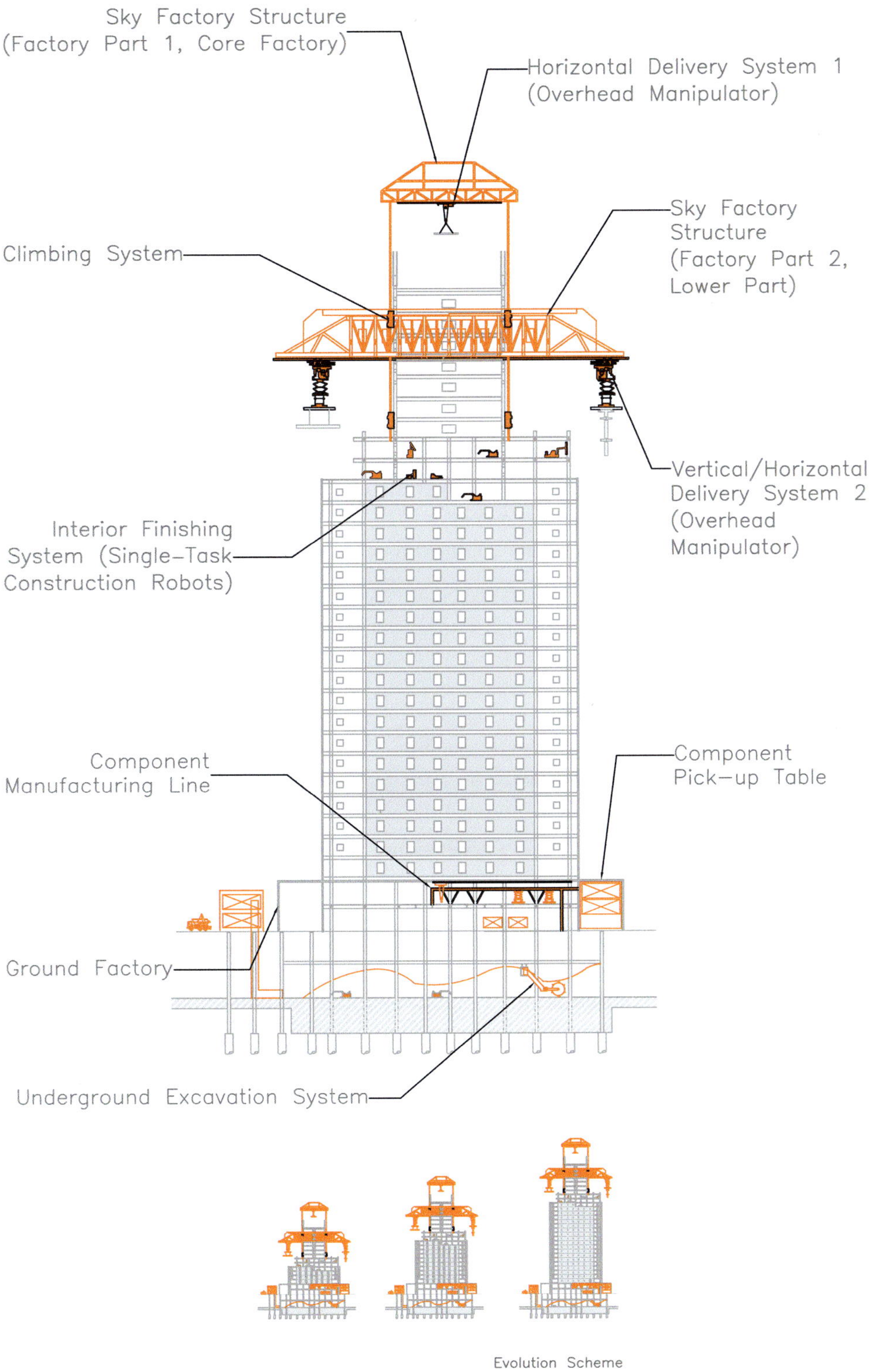

Totally Mechanized Construction System for High-rise Buildings (T-Up), Company: Taisei Corporation, Japan

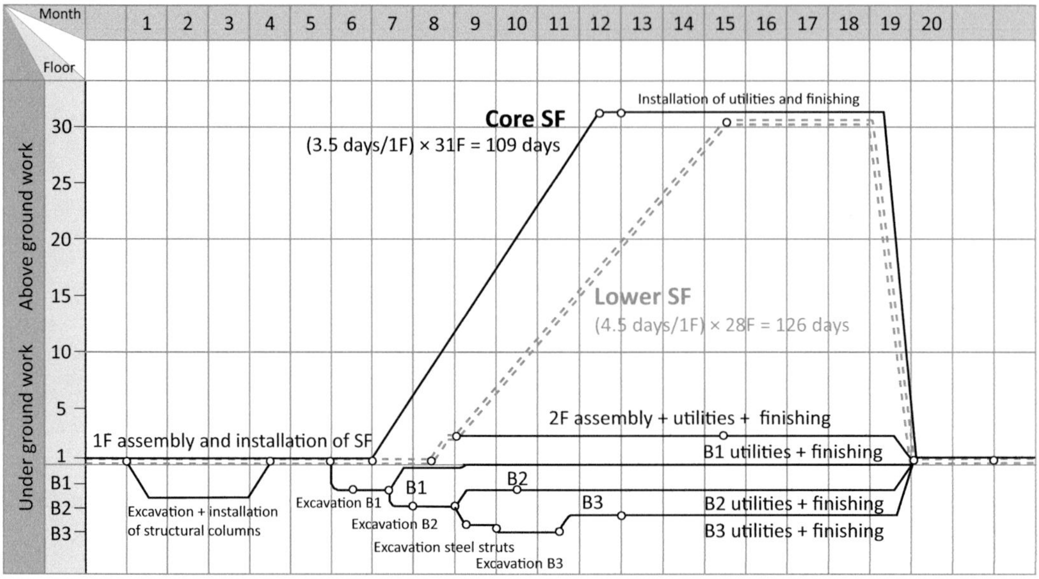

Figure 2.224. Overall schedule for the construction of a building by T-Up (adopted from Sakamoto & Mitsuoka, 1994).

FEC (Figure 2.225):

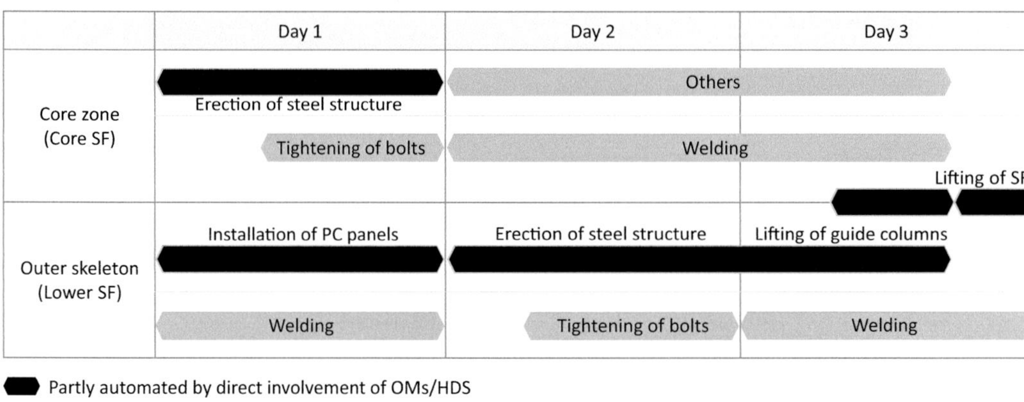

Figure 2.225. Floor erection cycle of T-Up (adopted from Sakamoto & Kumano, 1991).

Productivity (Figures 2.226–2.228):

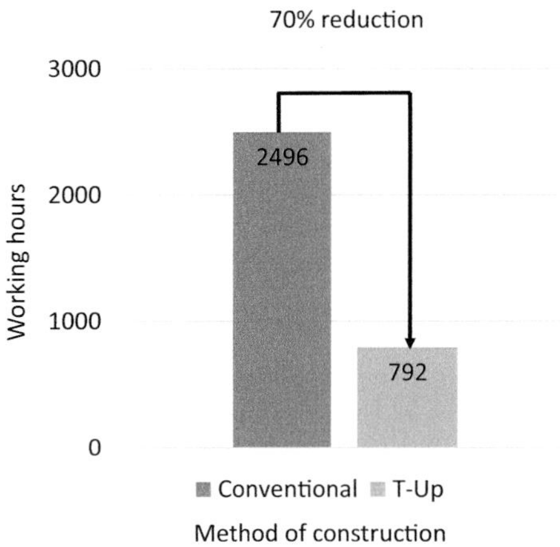

Figure 2.226. Comparison of labour productivity between conventional construction and T-Up application (adopted from documentary material T-Up/Taisei).

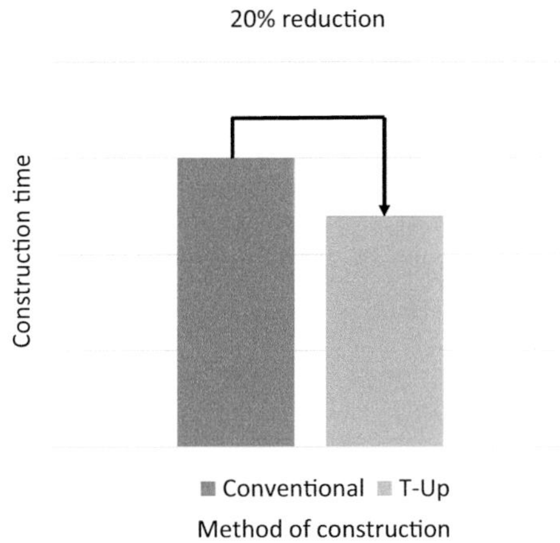

Figure 2.227. General reduction of construction time by T-Up compared to conventional construction (adopted from documentary material T-Up/Taisei).

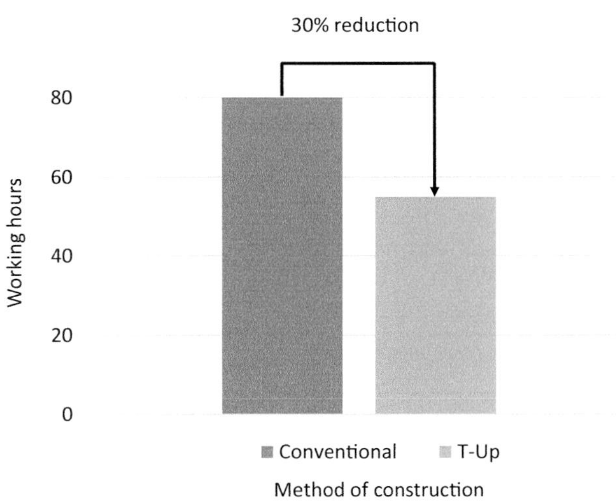

Figure 2.228. Reduction in time for removing the welding machines (adopted from documentary material T-Up/Taisei).

Figure 2.229. Sky factory in an advanced construction stage.

Figure 2.230. Finished product.

Figure 2.231. Sky factory with cantilevering parts for direct pick up.

Figure 2.232. View into sky factory part 1 and part 2. (Image: Taisei Corporation)

Figure 2.233. Ground factory.

Figure 2.234. Direct lift up of components.

Figure 2.235. View on cantilevering part of a sky factory.

Figure 2.236. Ground factory, gate.

Figure 2.237. Sky factory cantilevers for direct pick-up.

2.3.2 Robotic and Crane-Based Automatic Construction System (RCACS)

Company: Korean led consortium – KAIST, AIST (Japan); Doosan E&C, Ubiz Plus, Shinho System, CSSE, SMEC, M-Plus, Cospec, Robotous, Korea University, Yonsei University, Kunkuk University (Korea); S.I.T. (USA)
Category: SF pulled up by Core (main factory and core factory moving upwards)

Analysis of Technology and Composition

Evolution Scheme The RCACS (Figure 2.238) is developed for the on-site manufacturing of vertically oriented steel buildings with a concrete core. The core is first produced from precast concrete elements assembled by a V/HDS (robotic tower crane) and is then used to lift up an SF that provides an enclosed and SE for the assembly of the steel structure surrounding the concrete core. The assembly of the concrete core and the steel elements (in the SF) can be done in parallel. The V/HDS is used both for positioning of concrete elements and delivery/positioning of steel elements. The OM in the SF completes the process of component assembly (fixation). The system utilizes the Koreans' outstanding abilities in crane automation (V/HDS) and lightweight robot design (OMs).

Elevation In the monitoring and control system, all data from sensors deployed on the site are gathered through an integrated system protocol. Then, the real-time progress management (RTPM) and real-time visualization system (RTVS), the subsystems of the monitoring and control system, analyse the construction progress and visualize the status of the construction progress in 3D. By the V/HDS materials and structural steel frames are transported into the SF using a robotic tower crane installed within the concrete core of the building. An OM transports the bolting robot system (end-effector) to the working space in the SF, where it executes the bolting process.

Ground Plan The RCACS was intentionally developed and optimized for a certain type of building typology: high-rise buildings with a central core shaft. The whole assembly process is designed towards this feature: CS and V/HDS build on the structural core and therefore any kind of asymmetrical plan of the building is not suitable for the RCACS. The SF climbs while the building is assembled using the concrete core as vertical and structural support. The lightweight SF contains travelling rails for a lightweight, robotic gantry-type OM that moves radially around the core. So far, the system was only tested with one OM; also, the operation of several OMS in parallel would be feasible.

Subsystems The RCACS manufacturing the building on-site consists of a set of integrated subsystems:

1. *Sky factory structure*. The SF (Figure 2.239) is composed of structural frame modules fixed to the main structural core of the building. As this central core rises, the hydraulic jacks of the CS push up the factory. The SF integrates a roof and walls covered with sheets for providing a weatherproof and structured work area. The

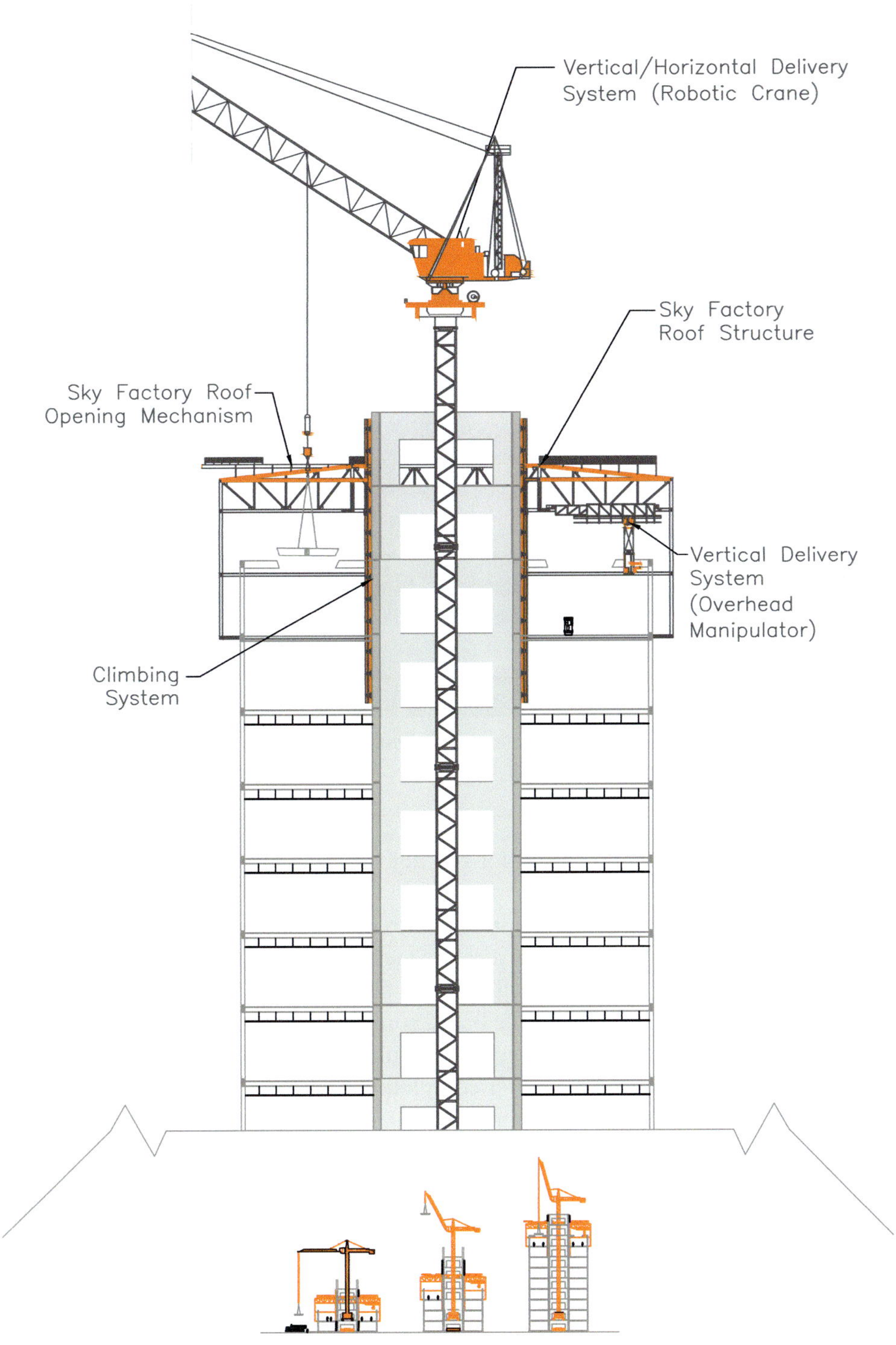

Robotic and Crane-based Automatic Construction System (RCACS), Company: Korean consortium

Figure 2.238. Horizontal delivery system/overhead manipulator.

roof can be opened and closed for the delivery of material into the SF by the robotic tower crane.

2. *Climbing system*. The construction factory is pushed up by an inchworm-like CS (Figures 2.246 and 2.247). The hydraulic mechanism works in two steps: in step 1 the mechanism pushes the factory up; in step 2 the mechanism climbs up the wall, sliding along the steel profile based consoles using an actuated system. The CS is a subsystem of the SF and contains eight hydraulic jacks, which push the SF farther up the structural core after completing each level. The maximum climbing speed equals to about 200 mm/min.
3. *Vertical/horizontal delivery system*. A robotic tower crane sits on top of the building's structural core and rises with the construction progress and the SF. The robotic tower crane is the main logistics system of the RCACS, used for supplying the construction site with materials and building components. It delivers components via the SF's automatically opening and closing roof sections. It can both directly position components (column components, beam components) and deliver the material for further manipulation and assembly by the robotic OM of the HDS. The robotic tower crane's operations are integrated with the material management system and the lifting operations planner.
4. *Horizontal delivery system*. The system has a total of 5 DOFs: 3 translations and 2 rotations. A central element is a robotic OM (Figures 2.240–2.244) that

allows various end-effectors (tools; Figures 2.248–2.250) to be driven into any desired position. A variation of the system can consist of a directly to the robotic girder added end-effector, or alternatively the existing bolting end-effector can be exchanged with end-effectors such as, for example, welding tools.

5. *Material management system*. The material supply system contains (operated from a control room; Figure 2.252) a complete database with all components and their properties modelled. The developers focussed on designing a construction progress visualization system by utilizing CAD and RFID technology. The system is called the RTPM (real-time progress management) system and is designed for monitoring the components, from their production in the off-site plant to the final assembly on the site. The material management system is integrated with the real-time visualization and control system (RTVS).
6. *Real-time visualization and control system*. Visualization and control module: HMI (human–machine interface) for emergency stops/other manual operations. It contains the following subsystems:
   - Lifting operations planner
   - Module for tracking the lifting operations and paths
   - Automated lifting module: monitors/adjusts differences between planned path and actual path of the robotic crane

End-effectors Various tools (end-effectors) can be mounted to the HDS (lightweight robotic overhead crane; Figure 2.241). The base component of the end-effectors is a cabin-like system (Figure 2.254) to which different tools (for example, for bolting; Figure 2.253) can be mounted. Different operations can be performed by changing either the tools mounted to the end-effector's cabin or by changing the cabin itself. The end-effector cabin is an exchangeable control cabin and tool carrier, positioned and orientated by the robotic OM, to which robotic tools such as a tool for positioning and bolting can be mounted. Further tools can be developed and attached to the cabin (e.g., for welding). In addition to the development of tools to be attached, new end-effector cabins can also be developed, for purposes such as component positioning or inspection.

System Variations In terms of efficiency, the robotic crane is not an optimal solution as it prevents, for example, extensive parallelization of logistics and delivery processes. In contrast to delivery systems of other systems it allows for great flexibility regarding building geometry and components sizes. However, this flexibility applies only for one, specific building typology. The RCACS is designed exclusively for the manufacturing of vertically oriented steel-based buildings with a central concrete core.

Robot-Oriented Design The beam components are designed on basis of ROD (Figure 2.251) to be guided into place automatically, therefore automating the process of adjusting after component positioning by the V/HDS. The tower crane places them into position (positioning and adjusting) and the bolting end-effector connects the components together (fixation). The beam components are split into lateral parts connected with the columns and junctions or intersections (cross-like shape) linking the lateral parts.

Figure 2.239. Sky factory, exterior view.

Figure 2.240. Overhead manipulator.

Figure 2.241. Sky factory, interior view.

Figure 2.242. Overhead manipulator, connector to the end-effector.

Figure 2.243. Overhead manipulator guided by the outer rails.

Figure 2.244. Overhead manipulator guided by the inner rails around corners.

Figure 2.245. View on overhead manipulator in action.

Figure 2.246. Actuator of the climbing system.

Figure 2.247. Climbing system.

Figure 2.248. End-effector.

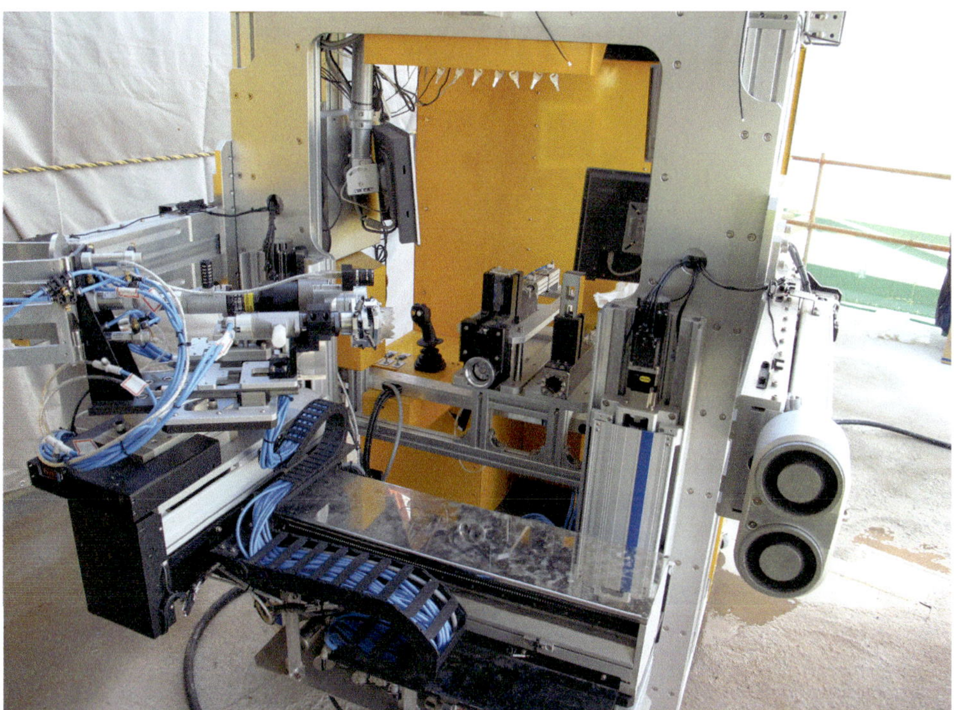

Figure 2.249. End-effector equipped with tools for automated screwing and bolting.

Figure 2.250. Control cabin, interior view.

Figure 2.251. End-effector for bolting. (Image: Korea University, 2010)

Figure 2.252. Control room. (Image: Korea University, 2010)

Figure 2.253. Installation of steel beam. (Image: Korea University, 2010)

Figure 2.254. Overhead manipulator equipped with end-effector for bolting. (Image: Korea University, 2010)

Figure 2.255. Horizontal delivery system/overhead manipulator.

2.4 Ground Factory (fixed location, vertically oriented building) and Building Push-up

Systems in this category are characterized by the installation of a fixed, ground-based, on-site factory for the erection of vertically oriented buildings. The main factory is located in the ground floors where the building's main structure (columns, beams, structural walls) is assembled. Systems in this category therefore utilize a rail-guided robot system on the ground, which picks up components from delivery trucks, storage, or the on-site sub-factory and positions them. The push-up of the assembled building structures is accomplished by hydraulic presses. After the push-up, the interior finishing on the upper floors is accomplished in parallel to the floorwise assembly of the main structure on the ground floor. The push-up mechanism simplifies assembly operation and logistical efforts as it allows major work to be done on the ground floor. Furthermore, the pushed up building provides a factory roof, taking away the need for a complex, temporary roof structure. However, because of a loading-bearing capacity limitation due to capacity limits of available and suitable push-up mechanisms, the height of the buildings that can be assembled is limited and it is not possible to construct super-high-rise buildings using the push-up mechanism.

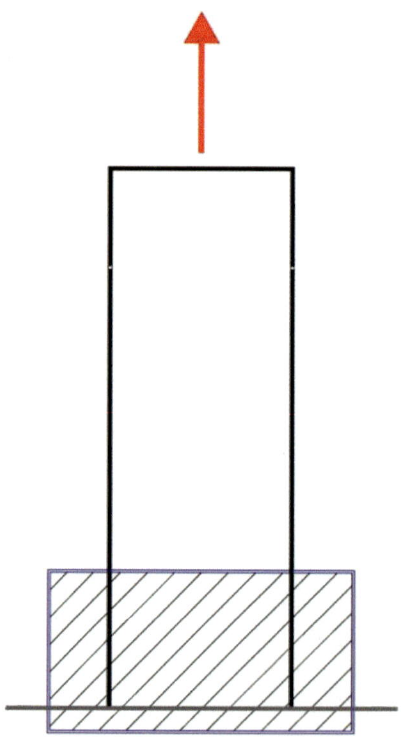

	System name	Company
13	Automatic Up-Rising Construction by Advanced Technique (AMURAD)	Kajima

Figure 2.256. Ground factory/rail-guided manipulators.

2.4.1 Automatic Up-Rising Construction by Advanced Technique (AMURAD)

Company: Kajima Corporation, Japan
Category: GF (fixed location, vertically oriented building) and building push-up

Analysis of Technology and Composition

Evolution Scheme In the case of the AMURAD, the building is assembled on the ground floor and pushed up floor by floor by screw presses. First, the screw presses and the rails required for the rail-guided, ground-based robots (Z-Carry main system; Figure 2.256) transporting and positioning prefabricated concrete components are installed. Then the GF structure covering all operations is erected.

Elevation The elevation shows the interplay of component positioning on the ground, the factory structure covering the lower floors in which the building is sealed, and the interior finishing work on the upper floors. Work is done on five levels in parallel; within these five working levels the building is completely finished. The first three levels are covered and enclosed by the factory structure. A representation of the work that could simultaneously take place over five floors in parallel can be seen below:

- Ground level (first floor): Assembly of main structural components
- Second level: Assembly of facade and balcony elements
- Third level: Finishing and sealing of facade
- Fourth level: Installation of interior walls, plumbing, appliances, toilets, and so forth (Figures 2.282 and 2.283)
- Fifth level: Final finishing, such as painting, and so forth

Ground Plan The ground plan shows the interplay of the rail-guided robot systems, the column grid, and the push-up mechanism on the ground level. Each column is assigned a push-up mechanism. The beams span the whole width of the building. The components processed are reinforced concrete components (main elements of the kit: columns, beams, floor slabs, facade elements, balcony elements). The prefabricated concrete elements are joined and fixed with *in situ* concrete on the ground level on-site. The push-up mechanism performs lifting so accurate that it does not affect the hardening process negatively (Figure 2.290).

Subsystems The AMURAD manufacturing the building on-site consists of a set of integrated subsystems:

1. *Ground factory structure.* The GF structure provides a covered and structured work environment for the first three floors. Within these three floors (Figures 2.263–2.265, 2.284, 2.286, and 2.289), the main structure, facade elements, and balcony elements are assembled, and the building is sealed. Furthermore, the factory structure provides working platforms for facade work as well as a frame in which a VDS can be installed.
2. *Push-up mechanism (referred to by Kajima as the Z-Up system).* A screw press is installed under each column (Figures 2.266 and 2.267). The screw press functions as jig/fixture that guides the column positioned by the Z-Carry main robot into place (Figures 2.268–2.271). In addition, after positioning, alignment, and fixation of all the columns and beams on the ground level, the screw press smoothly and precisely pushes up the building so that the next floor can be erected on ground level.
3. *Rail-guided ground-based main robot (referred to by Kajima as the Z-Carry main system).* A rail system is employed on the ground level for the picking up and positioning of columns, beams and floor slabs by a robot (Figures 2.271, 2.273, and 2.274). The robot has 4 DOFs (translation on rails lengthwise, translation sideways, rotation, lowering/lifting of components). It can be equipped with component-adjusted end-effectors with additional DOFs (e.g., a floor slab installation end-effector adds an additional DOF).
4. *Rail-guided ground-based horizontal logistics assistant robots (referred to by Kajima as the Z-Carry subsystem).* On the ground, a rail system for two horizontal logistics robots (Figures 2.272, 2.275, 2.284, 2.285, and 2.288) is installed flanking the main robot on both sides. The two robots transport beam elements and floor slabs into the factory and supply the main robot. The main robot can thus operate only in the factory and does not have to go back and forth to pick up components.
5. *Material handling, sorting, and processing yard (optional).* An MHSPY can be erected on the ground level into which a rail-guided main robot and horizontal logistics assistant robots can drive. An overhead crane positions components onto the robot systems.
6. *Vertical delivery system (referred to by Kajima as the Mighty Hand).* A rail-guided robot delivers (within the GF structure) material to the second and third floors for interior finishing (Figures 2.294–2.301). The robot system can also deliver

loaded mobile and immobile parts pallets (Figures 2.278–2.280) as a mechanism that allows the material to remain balanced in an upright position during transport. The robot can rotate its final link, on which the material or roll car is positioned, and position it on the second and third floors in a way that is easy to deliver delivery or pick up by workers.

7. *Balcony positioning system.* A balcony position system works on a rail system on the third floor of the factory structure, transporting and positioning prefabricated balcony components (Figure 2.281).
8. *On-site component storage and MHSP (optional).* Apart from the GF, sub-factories (Figures 2.293) can be erected that can function as storage for the on-site factory and as covered processing yard for the structured on-site transformation of low-level into high-level components (e.g., on-site production of the prefabricated concrete components; Figure 2.292). The VDS can be used to transport material/components JIT to the second and third floors via a rail system that extends into the sub-factory (Figures 2.295–2.297).
9. *Real-time monitoring and management system.* Data from sensor systems as well as from the servomotors/encoders can be read to control equipment operation and to monitor the component assembly process from a control room (Figure 2.276).

End-effectors The systems described in the preceding text can be equipped with various end-effectors to suit the picking up, positioning, and adjustment of various component types. The speciality of the AMURAD is that the main manipulating system (Z-Carry) works not only with component adjusted end-effectors, but also with various other subsystems:

1. End-effector for pickup, positioning of columns and beams (Z-Carry main system end-effector; Figure 2.271)
2. End-effector fixture in which the column is positioned and the automatically adjusted (Z-Up End-effector; Figure 2.268)
3. End-effector for transport of beams (Z-Carry subsystem end-effector)
4. End-effector for transport of floor slabs (Z-Carry subsystem end-effector; Figure 2.285)
5. End-effector for pickup, positioning of floor slabs (Z-Carry main system end-effector; Figure 2.277)
6. End-effector for delivering carriers with materials for interior finishing to the second/third floors (Mighty Hand end-effector; Figure 2.278)
7. End-effector for positioning of balcony elements (end-effector of balcony positioning system; Figure 2.281)

System Variations The size of the building can be varied lengthwise as long as the column grid structure is synchronized with the positioning of the push-up mechanisms. The minimum width depends on the accessibility by the Z-Carry's rail-guided main and sub robots, and the maximum width on the ability of the beam elements to span the distance between the structural elements. Basically, the arrangement of the components should follow an orthogonal arrangement. However, the columns at the end of each front side can be irregularly arranged, and the rail system can also follow a curve (Figure 2.291). Furthermore, components can be created from parts and low-level components off-site or on-site in a sub-factory. Another scenario envisioned by Kajima was for the system to be used to enhance existing buildings or to integrate damping systems (on the ground level) into existing buildings. To achieve this, the push-up mechanisms (Z-Up) are installed next to existing columns; the building is cut at the ground level and pushed up slightly and then held in position by the screw presses' end-effectors. From then on the above described process of lifting and component (or damping system) installation on ground level can start.

Robot-Oriented Design Columns are designed (Figures 2.271 and 2.287) to simplify the operation of the rail-guided ground-based main robot (Z-Carry) and its end-effectors. When considering the beam and floor slab components, the dimensions and potential weight of the main robot is taken into account. On a building and urban level the orientation of the building and thus the main logistics and main access port for the ground-based main robot system has to be considered in the planning phase. The urban surrounding also determines if optional elements as the sub-factories could be installed on the long side of the building.

Efficiency Analysis

Kajima uses its AMURAD system to erect concrete-based buildings with up to 20 floors. In contrast to other systems, AMURAD installs the SF on the ground. Rail-guided robots position the concrete elements on the ground and hydraulic presses push up the completed floors. In parallel, interior, plumbing, and finishing work is done on the upper floors.

Detailed Project Schedule See Figure 2.257.

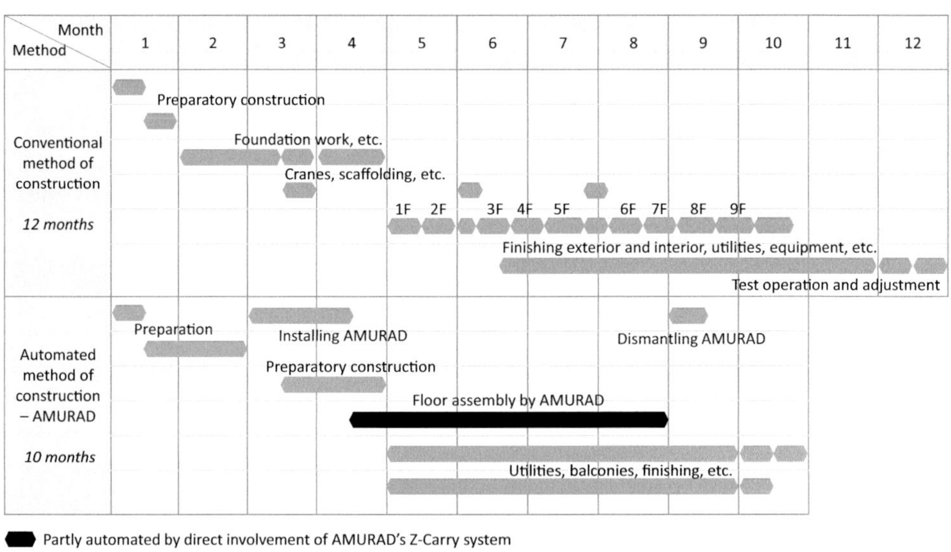

Figure 2.257. Comparison between conventional construction and automated construction (AMURAD) (adopted from documentary material/Kajima).

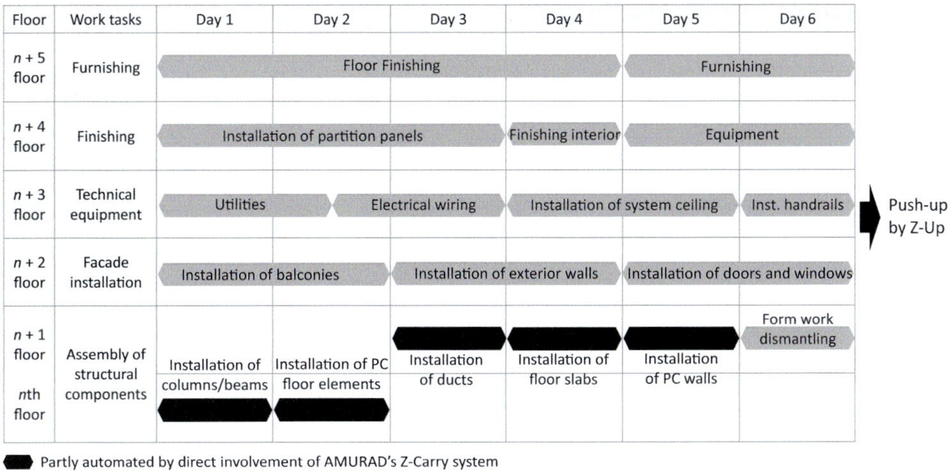

Figure 2.258. AMURAD's six day floor erection cycle (adopted from Kajima, 2000).

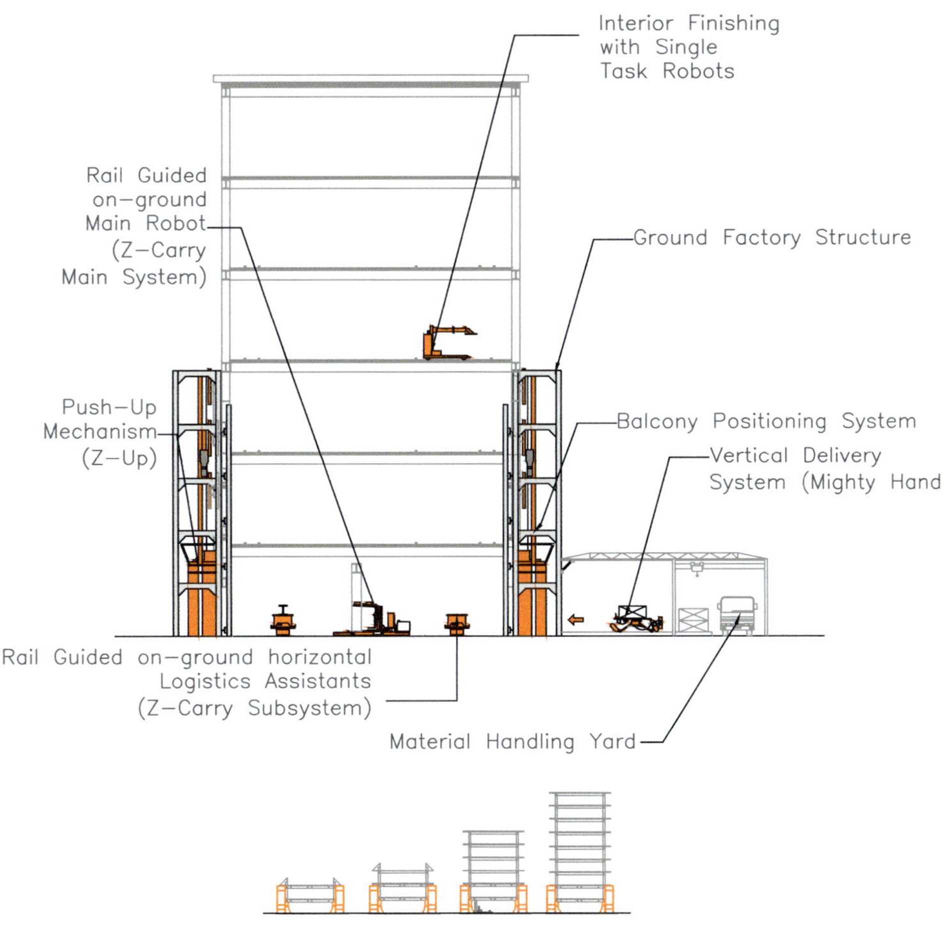

Automatic Up-Rising Construction by Advanced Technique (AMURAD), Company: Kajima Corporation, Japan

Table 2.6. *Specifications of Z-Up, Z-Hand, and Z-Carry*

Item	Specifications of Z-Up	Item	Specifications of Z-Hand	Item	Specifications of Z-Carry
Ability to push up	400–600 tons	Weight of loading	5 tons	Ability to push up	1.3 tons
Outline dimensions	B2.700×W4.000 ×H6.000	Outline dimensions	L5.500×W3.500 ×H2.425	Outline dimensions	L4.500×W1.400 ×H1.660
Weight of equipment	About 50 tons	Weight of equipment	About 8 tons	Weight of equipment	About 5 tons
Climbing system	Electric clinchers and nuts	Ascent stroke	2.069 m	Moving velocity	Up to 10 m/min
Stroke	3.250–3.950 mm	Moving velocity	15 m/min	Operating method	Automatic moving
Ascent velocity	15 mm/min 40 mm/min (without load)	Operating method	Wireless remote control joystick	Technology	Inverter
Tolerance	0.1 mm	Moving method	Rail	Moving method	Rail 3D monorail
Engine	AC 400 V, 10 pieces of equipment use 254 kW	Engine	AC 400 V, (30 kW each)	Engine	AC 400 V, (30 kW each)

Adapted from Hasegawa (1999).

Floor Erection Cycle During the first applications it was possible to push up one floor per six working days and thus to conduct assembly and interior finishing work on the basis of a six-day cycle time (Figure 2.258). However, it has to be considered that the system was still in a developmental phase.

Technical Data, Speed of Equipment The key subsystems (see also Table 2.6) of the AMURAD can be operated both manually (remote controlled) and automatically. The maximum operation speed of the subsystems:

- Z-Up: 10 m/min
- Z-Carry: 10 m/min
- Z-Hand: 10 m/min

Productivity The AMURAD (Figures 2.259 and 2.260) was among the first projects able to improve work productivity by more than 20% (measured by man-days/m^2)

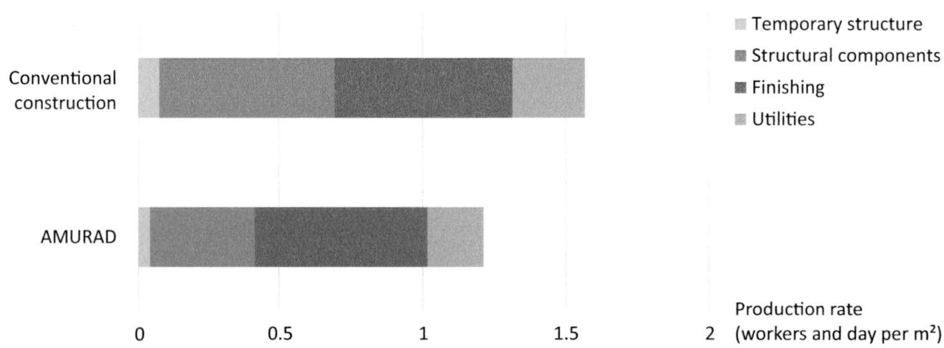

Figure 2.259. Comparison of construction work production rate between conventional construction and AMURAD (adopted from Sekiguchi et al., 1997).

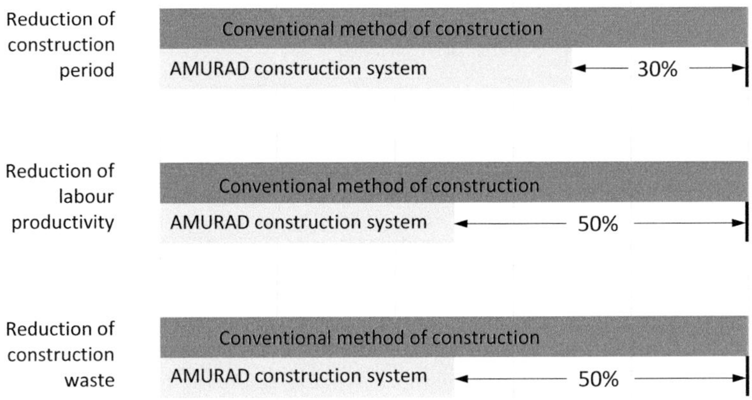

Figure 2.260. Comparison of construction time, labour productivity, and construction waste generation between conventional construction and AMURAD (adopted from Kajima, 2000).

compared to conventional construction of comparable buildings. The highest reduction of human labour (nearly 50%) was in relation to the erection of the building's concrete structure (structural member).

Learning Effects Kajima measured the time needed for the installation of components in a building with nine floors over eight cycles. In general, the time needed could be reduced (learning effect) in all installation tasks by 40–50% (Figure 2.261). For example:

- Installation of precast concrete columns by Z-Carry: Reduction from 20 minutes to 13 minutes
- Installation of precast concrete girders by Z-Carry: Reduction from 20 minutes to 9 minutes
- Installation of precast concrete floor slabs by Z-Carry: Reduction from 11 minutes to 6 minutes

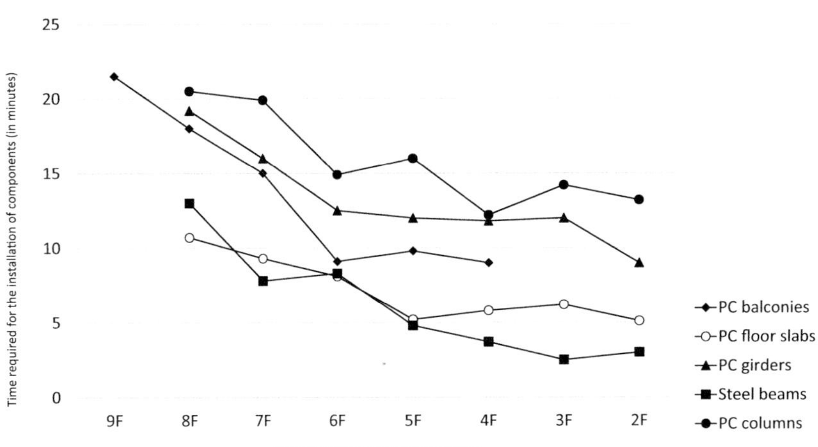

Figure 2.261. Learning effects and time required for installation of various components (adopted from Sekiguchi et al., 1997).

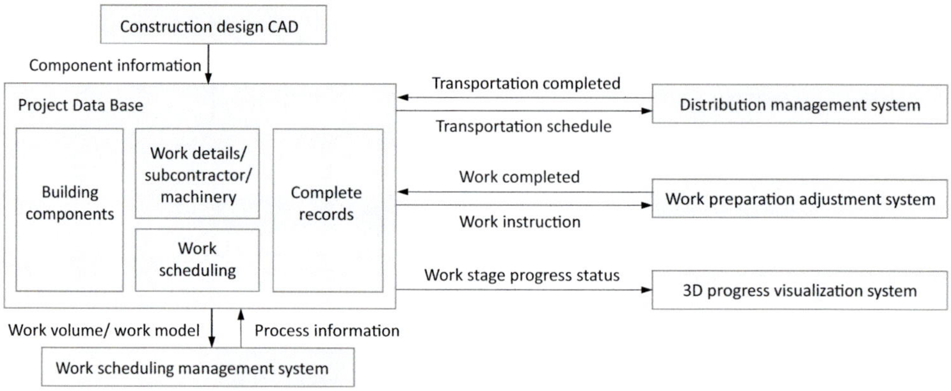

Figure 2.262. Information integration and management scheme (adopted from Sekiguchi et al., 1997).

Product Quality and Process Monitoring AMURAD is based on an "information integration system" (Figure 2.262) that helps to set up, control, and supervise the construction process. The system allows the optimization and control of material flow, displaying the work progress status and control time and quality. The system is able to partly automate the generation of work schedules and operation instructions for equipment operators and workers.

Figure 2.263. System set-up.

Figure 2.264. System has produced several floors.

Figure 2.265. Ground factory, side view.

Figure 2.266. Setting up of screw presses on the ground level (Z-Up system).

Figure 2.267. Screw press.

Figure 2.268. Screw presses (Kajima, 2000).

Figure 2.269. Screw press end-effector fixture (Kajima, 2000).

Figure 2.270. Push-up mechanism (Z-Up system) (Image: Kajima, 2000).

Figure 2.271. Rail-guided ground-based main robot (Z-Carry main system) (Image: Kajima, 2000).

Figure 2.272. Rail-guided ground-based horizontal logistics assistant robot (Z-Carry subsystem). (Image: Kajima, 2000)

Figure 2.273. Rail-guided ground-based main robot (Z-Carry main system). (Image: Kajima, 2000)

Figure 2.274. Rail-guided ground-based main robot (Z-Carry main system). (Image: Kajima, 2000)

Figure 2.275. Rail-guided ground-based horizontal logistics assistant robot (Z-Carry subsystem). (Image: Kajima, 2000)

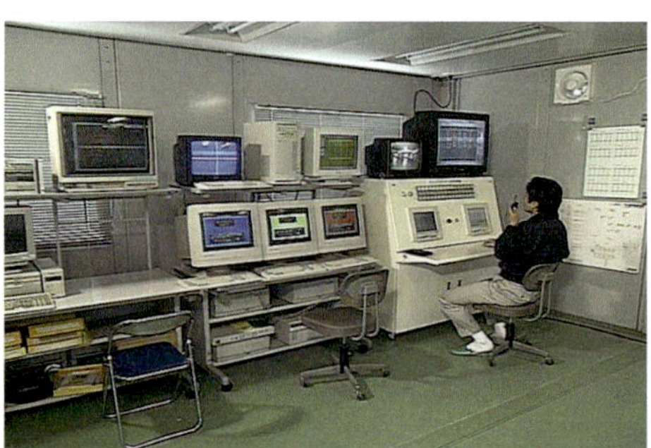

Figure 2.276. On-site control room. (Image: Kajima, 2000)

Figure 2.277. Rail-guided ground-based main robot (Z-Carry main system). (Image: Kajima, 2000)

Figure 2.278. VDS (Z-Hand). (Image: Kajima, 2000)

Figure 2.279. VDS (Z-Hand). (Image: Kajima, 2000)

Figure 2.280. On-site material management. (Image: Kajima, 2000)

Figure 2.281. Balcony positioning system. (Image: Kajima, 2000)

Figure 2.282. Parallel interior finishing on the upper levels. (Image: Kajima, 2000)

Figure 2.283. Interior finishing using prefabricated bath modules (high-level components).

Figure 2.284. Ground factory, rail-guided ground-based main robot Z-Carry main system (middle) and automated logistics assistants of the Z-Carry subsystem (right and left).

Figure 2.285. Ground-based and rail-guided automated logistics assistant.

Figure 2.286. Rail-guided and ground-based main robot, detailed view.

Figure 2.287. Component design synchronized with main robot's end-effector.

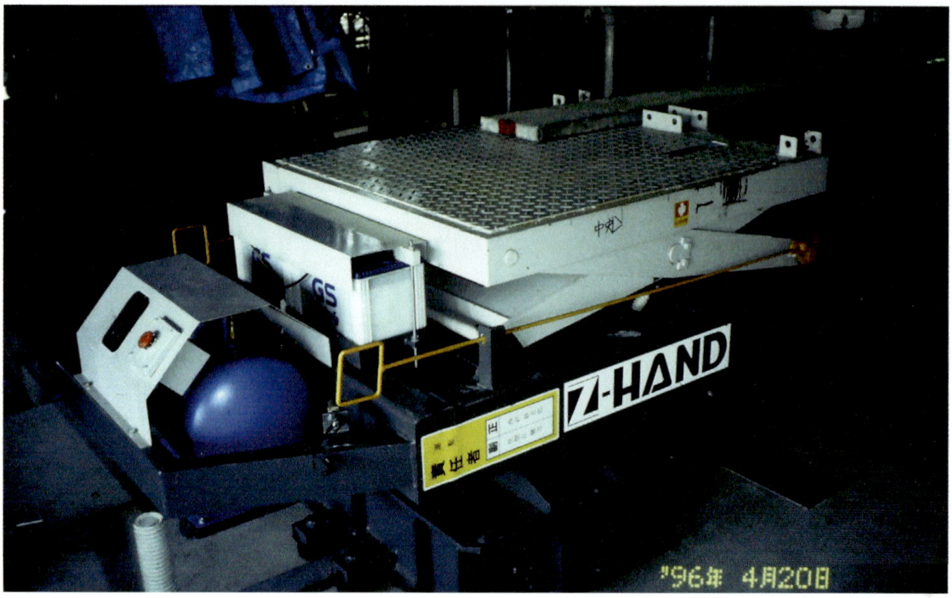

Figure 2.288. Ground-based and rail guided automated logistics assistant, detailed view.

Figure 2.289. Trucks delivering material just in time just in sequence for direct pickup in a sub-factory.

Figure 2.290. On-site concreting of connections between concrete components.

Figure 2.291. Rail system can be adjusted for not-rectangular buildings.

On-site Sub-factories

Figure 2.292. On-site production of precast concrete components.

Figure 2.293. Sub-factory, outside view.

Figure 2.294. A rail system connects sub-factory and main factory.

Figure 2.295. A rail system connects sub-factory and main factory.

Figure 2.296. Z-Hand moving on rails.

Figure 2.297. Horizontal delivery system (Z-Hand) loaded with material in the sub-factory.

Figure 2.298. Vertical delivery system (Z-Hand) transporting material for interior finishing to upper floors.

Figure 2.299. Z-Hand control box in the sub-factory.

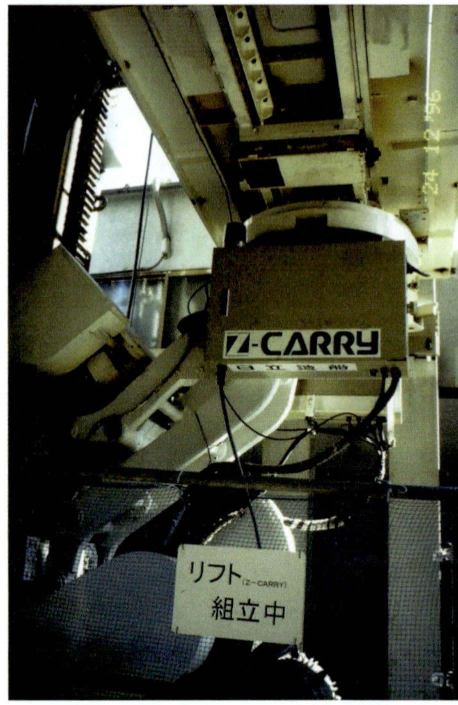

Figure 2.300. End-effector of the Z-Hand.

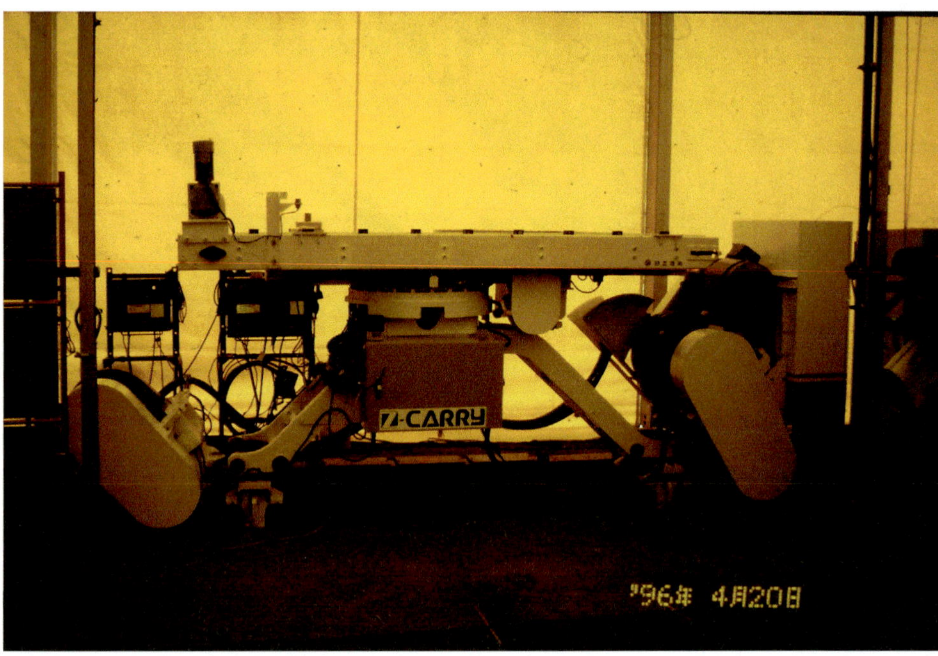

Figure 2.301. The Z-Hand robot represents the main element of the vertical delivery system.

2.5 Ground Factory (fixed location, horizontally oriented building) and Building Push-up

Systems in this category are characterized by the installation of a fixed, ground-based on-site factory for the erection of horizontally oriented buildings that are long and/or wide horizontally, while at the same time relatively flat. Skanska, unlike Kajima, developed its system not for the construction of vertically oriented buildings, but for horizontally oriented condominiums. Also, in terms of component size and complexity, it takes a different approach. Whereas the positioning robot of the AMURAD system processes a multitude of smaller and lighter components, Skanska's positioning robot processes a smaller amount of larger and heavier wall elements. A similarity between the AMURAD system and System Skanska is the installation of a ground-based, rail-guided, robot system.

Sekisui Heim's J-Up system could also be placed in this category. The J-Up system (so far) automates only the Push-up mechanism. The assembly of the prefabricated panels on the ground level, as well as the interior finishing, are done manually. The system allows hydraulic jacks to be placed freely, and a complex rail system to guide robots is not needed. The system thus allows for a high degree of flexibility in terms of the layout and shape of the building, as well as in particular its horizontal extension. Owing to the characteristic of combining automated systems with manual work in parallel to hybrid systems, relations to other categories exists (centralized SF in combination with conventional construction; decentralized SF in combination with conventional construction).

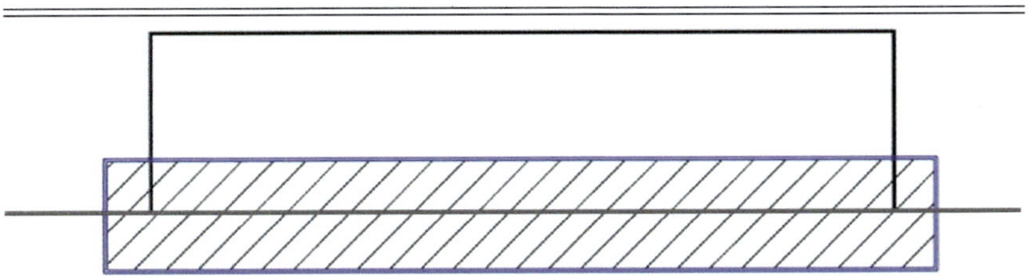

	System name	Company
14	System Skanska	Skanska
15	J-UP	Sekisui House

However, owing to the location and working-direction oriented focus of this categorization, the system was allocated to this category (GF – fixed location, horizontally oriented building – and building push-up). In contrast to other systems in this and the preceding category, the J-Up is a relatively low cost system and thus lowers the barrier-to-entry. It was used by Sekisui House itself as well as by subcontractors who did the on-site assembly work. The push-up methods developed (especially those simpler ones of J-Up and System Skanska, but also the AMURAD from former category) follow the traditional construction of timber buildings in Japan where the complex roof structure is constructed first on the ground and then pushed up, providing a shelter for completion of the building.

2.5.1 System Skanska

Company: Skanska Group, Sweden
Category: GF (fixed location, horizontally oriented building) and building push-up

Analysis of Technology and Composition

Evolution Scheme The Swedish construction group Skanska has developed this system for the automated construction of horizontally oriented buildings (such as, e.g., condominiums) which are the predominant building typology in Europe. The development of the system was inspired by the approach of Kajima's AMURAD system and utilizes a fixed GF in which rail-guided and ground-based robots operate in a structured environment. In contrast to the AMURAD system, operations are more horizontally oriented and the building is built up from prefabricated wall panels and not from columns and beams. The evolution sequence can be characterized as follows.

1. Installation of GF and ground-based robots/HDS
2. Floor assembly on the ground
3. Pushing up of the assembled floor
4. Repetition of 2. and 3. and parallel interior finishing

Figure 2.302. Factory, horizontal delivery system (rail-guided manipulator). (Image: Skanska)

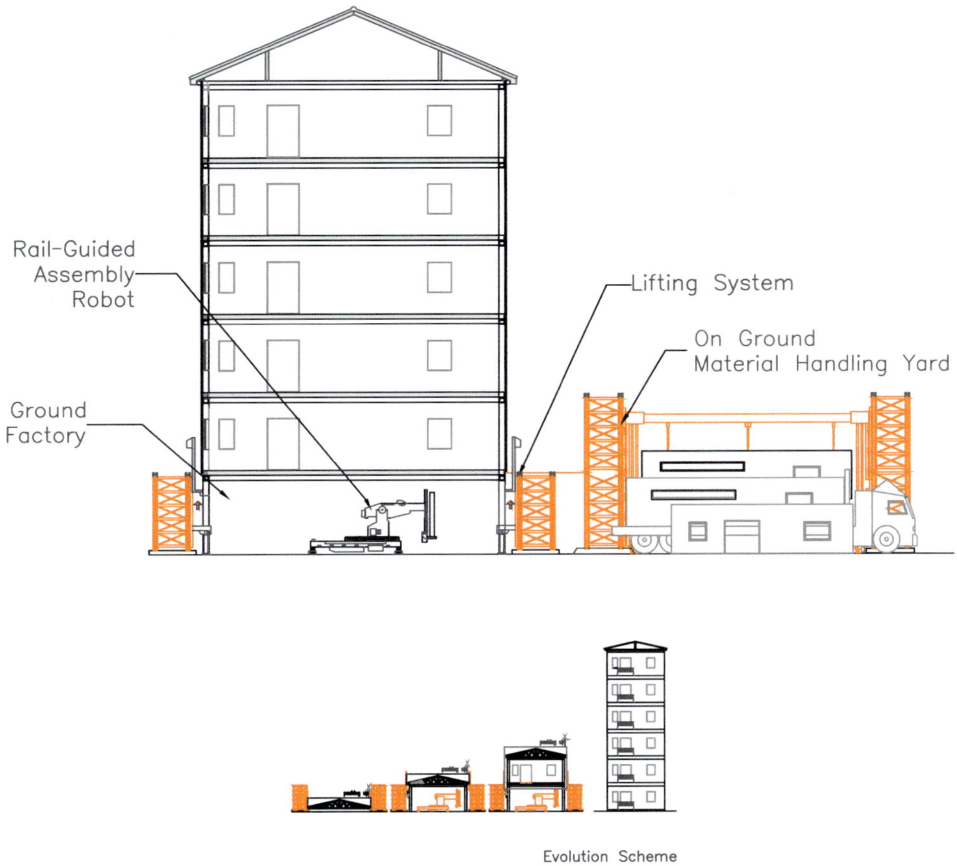

System Skanska, Company: Skanska Group, Sweden

Skanska has applied modularization strategies to the building-component system:

- Utilization of 80% standardized concrete components that can be handled by the HDS on the ground with a reduced level of complexity.
- Usage of 20% building-individual and customized parts and lower-level components (according to Skanska this is sufficient to deliver a high-value unique product to the customer at a reasonable cost).

Elevation The positioning and alignment of the structural main elements (concrete components) is done on the ground floor by the ground-based HDS. Further work is done in parallel on the ground floor and upper floors. The detailed work process sequence can be described as follows:

1. Factory external logistics (transport of components to the site)
2. Use of unloading system for unloading wall components form the truck delivering those to the site
3. Temporary storage of the wall components

4. Assembly robotic system collects the wall components from temporary storage and carries them to the on-site factory
5. Top floor assembly on ground
6. Pushing up of the first floor
7. Repeat this process again and again until the completion of the building

Ground Plan The construction site is overlaid with a grid that is in tune with the building's structure. Guiding rails for the HDS are then installed on this grid for the operation of the assembly robots. Then the lifting system is placed on the grid as well. Similarly, material handling systems and material stocks completing the on-site factory are allocated.

Subsystems Skanska's integrated automated building construction site, consists of many integrated subsystems:

1. *Ground factory*. Steel structures and lifters are used as temporary load bearing structures during the construction of the building. After completion of the building, those are removed. The GF structure (Figure 2.302) can optionally be covered, thus creating a weather-protected work environment.
2. *Lifting system*. After completion of a floor on the ground, heavy-duty hydraulic cylinders are used to push the whole floor up onto the temporary load bearing structures (Figures 2.303, 2.306).
3. *Ground-based HDS (rail-guided assembly robot)*. Picking up the modules and then delivering them to the construction area, where they are integrated into the building (Figures 2.304, 2.305). The system is able to position and orientate components.
4. *Ground-based MHSPY (optional)*. A rail-mounted overhead gantry crane can be installed for the unloading of concrete components from trucks delivering them to the site.
5. *Material unloading automated system (optional)*. An automated material unloading system is used to unload the modules/components from the truck delivering them to the site.
6. *Controlling systems*. This lifting process must be very strictly controlled and thus the hydraulic cylinders are fitted with an internal position transducer. Similarly, the pressure lines have pressure transducers, all being located internally such that they are protected from inclement weather, dirt, humidity, and so forth. All the information is received at a control panel, which, by means of a programmable logic controller (PLC), manages the data and sends orders to the electro valves. A control panel allows the operators to monitor the load and position of each of the cylinders continuously and in real time, and to stop the lifting operation if any problem occurs. Furthermore, oil levels and temperature are monitored and the operation stops automatically in case of certain deviations.
7. *Supply system*. Various precast concrete prefabrication plants operated by Skanska can be integrated with the management and control system, and can deliver components JIT, JIS to the site.

End-effectors Ground-based HDS (rail-guided assembly robot) can be equipped with various end-effectors, necessary for picking up concrete components. Owing to the high-degree of standardization of the concrete components, so far only one end-effector has been required. However, if greater variety is required, additional exchangeable end-effectors can be developed.

System Variations Owing to the modular approach followed by Skanska, the following variations to the system are possible:

- *Using more than one assembly robot.* The system in its present developmental stage uses only one ground-based assembly robot system. However, in case several ground-based assembly robot systems are used in parallel to produce different sections of the building, then the time required for manufacturing of the building can be considerably reduced.
- *Parallel construction in more than one rail track.* In its present form the system focuses on a single track for the operation of the ground-based robots. However, by using several rail tracks and robots larger buildings such as schools or hospitals can also be built by this system efficiently.

Robot-Oriented Design: On a component level, ROD is employed to design the prefabricated wall elements in a way (geometry, size, weight) that they can efficiently be handled by the ground-based robot system/HDS. On a building level, ROD is applied in the sense of keeping the level of standardized building components that can be assembled by the robot system with 80% as high as possible to foster automation. On an urban level, grid-based and modular planning is applied and the allocation of material delivery and handling yards in combination with the site layout can optimize the manufacturing procedures.

Figure 2.303. Lifting after assembly of ground floor. (Image: Skanska)

Figure 2.304. Horizontal delivery system. (Image: Skanska)

Figure 2.305. Mobile control interface. (Image: Skanska)

Figure 2.306. Lifting system. (Image: Skanska)

2.5.2 J-Up

Company: Sekisui House, Ltd., Japan
Category: GF (fixed location, horizontally oriented building) and building push-up

Analysis of Technology and Composition

Evolution Scheme The basic concept of the J-Up method (Figure 2.307) is to build the upper floor first, and then finish the assembly of the lower floors under the lifted uppermost floor in a partly weatherproof and structured work area. Sekisui House used mobile hydraulic jacks to make the concept possible and it prefabricates and prepares in an orderly way all elements in its highly automated off-site factories (Figure 2.333; see **Volume 2** for further information). First, the hydraulic jacks are located on the finished foundation and the roof is assembled directly on them. Second, the jacks lift the roof up and the workers finish assembling the walls of the second floor. Third, the jacks lift the finished second floor up, and the workers finish assembling the walls of the first floor in an SE. Finally, workers install the interior fittings and prepare the house for handing over to the client.

Elevation At first, the foundation is prepared (Figure 2.308). J-Up hydraulic jacks are positioned. Then, the roof and the associated floor beams are assembled (Figure 2.309); special supporting steel frames are put in place under the roof and over the foundation (to support, for example, in case of an earthquake during the lifting up process). After this, the next period of lifting up is conducted. The prefabricated and to the site delivered steel frames are then assembled between the roof and the floor beams level (Figures 2.310 and 2.311), the supporting frames are replaced with prefabricated steel frames, and assembly of the insulation and facade for the total floor is finished. At this stage the second floor is thus completed and can be lifted up as well (Figure 2.312).

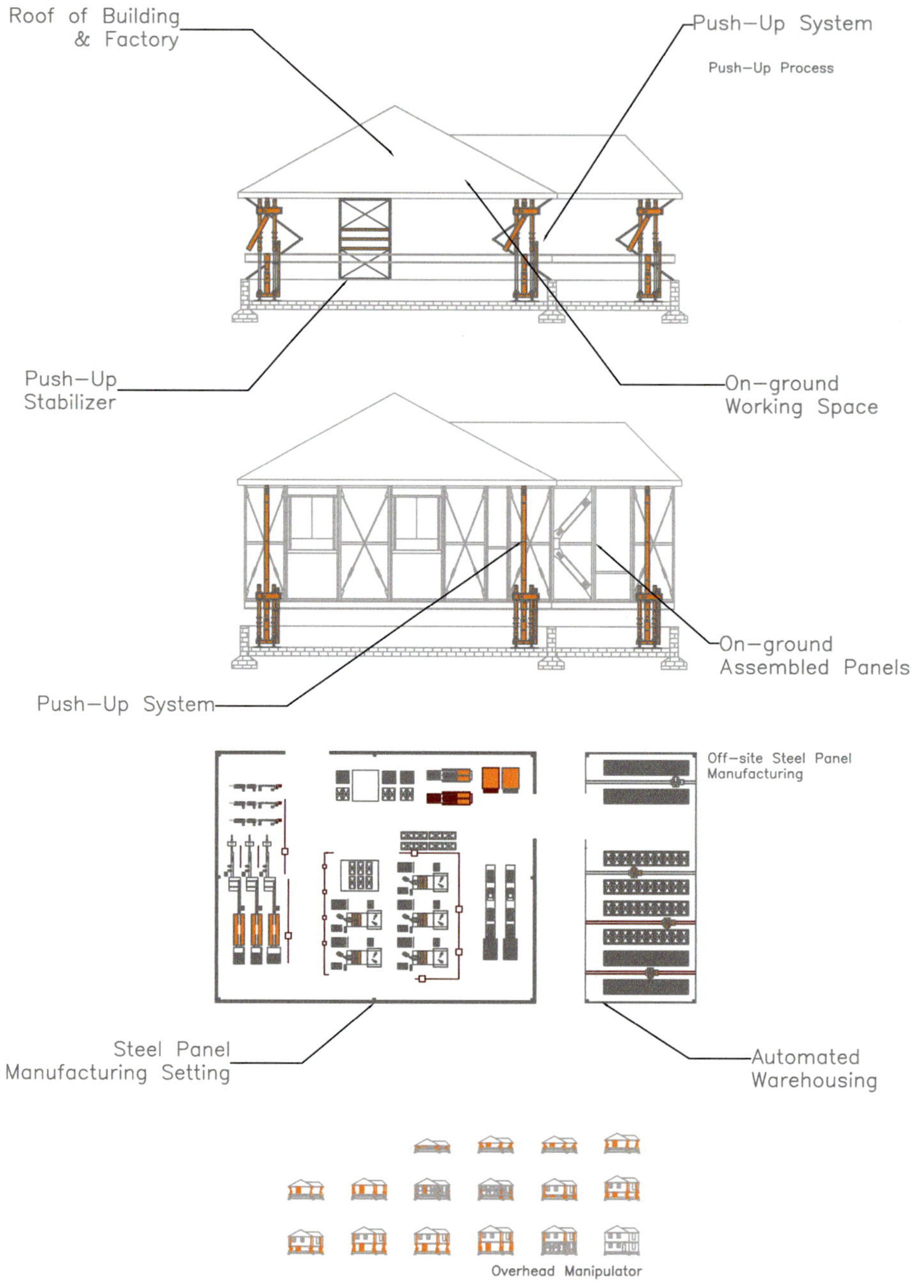

J-Up, Company: Sekisui House, Ltd., Japan

The special supporting steel frames are again positioned under the finished second floor and fixed to the foundation and the next period of lifting up is commenced, and the ground floor (first floor) is assembled and finished (Figures 2.313, 2.315, 2.318, and 2.319).

Ground Plan J-Up hydraulic jacks are located at fixed positions determined by force analysis software, which is based on the floor plan layout. Principally, Sekisui House uses 6–12 jacks on one site. The J-Up jack is a twin-cylinder hydraulic jack whose maximum load is 4 tons. The lifting speed of the jack is 900 millimetres in

Figure 2.307. Hydraulic jacks lift up finished floors and allow convenient assembly of the factory prefabricated elements on the ground.

approximately –5 minutes. With the help of J-Up jacks, the largest floor area that can be constructed is about 250 square metres and two to three floors at most.

Subsystems The J-Up (system) consists of integrated subsystems:

1. *Hydraulic jacks*. Computer-controlled jacks are used to push up the building (Figures 2.314, 2.316, 2.317, 2.320, and 2.321). Stilts segments are inserted into the jacks during the pushing up process to extend the push-up stilts on which the upper part of the building is temporarily resting.
2. *Supporting steel frames*. Special supporting steel frames are positioned temporarily under the finished and pushed up floors and are fixed on the foundation so that a collapse of the building in case of an earthquake is prevented (Figure 2.319).

System Variations The system is able to erect any building based on Sekisui House steel frames. The number of jacks can be varied depending on the size and shape of the building.

Figure 2.308. Preparation of foundation and jack system.

Figure 2.309. Assembly of roof.

Figure 2.310. Pushing up of roof.

Figure 2.311. Ground-based assembly of upper floor/floors.

Figure 2.312. Pushing up upper floor/floors.

Figure 2.313. Assembly of ground floor.

Figure 2.314. Push-up mechanism (hydraulic jacks).

Figure 2.315. Assembly of prefabricated steel frames.

Figure 2.316. Push-up mechanism (hydraulic jacks).

Figure 2.317. Push-up mechanism (hydraulic jacks).

Figure 2.318. Assembly of prefabricated steel frames.

Figure 2.319. Assembly of prefabricated steel frames.

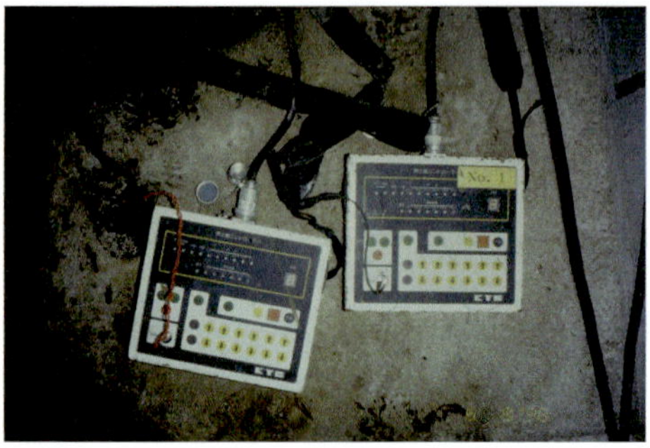

Figure 2.320. Control system of the push-up system.

Figure 2.321. Control system of the push-up system.

Figure 2.322. View onto the production line of a factory of Sekisui House. (Photo: Sekisui House)

2.6 Off- and On-site Combined Factory (both in a fixed location)

Systems in this category are characterized by the close cooperation of an off-site factory and an on-site factory, both of which are operated by the building manufacturer. The on-site factory is a modular and temporary hall structure completely independent from the building. It encompasses the building completely and is self-supporting. The components delivered to the site are high-level components and delivery and installation are organized strictly JIT and JIS. The temporary on-site hall provides a sealed and heated work environment as well as an overhead crane system. From 2005 to 2007 NCC ran a test project for a system falling under this category: NCC Komplett. The NCC Komplett system was a manufacturing system fully integrating factory prefabrication with a mobile on-site assembly hall. In the fixed off-site factory; concrete walls were customized according to the architect's plans; and even installations such as windows, doors and radiators were integrated. In the factory, the high-level wall and ceiling components were completely preinstalled and finished, which meant that even furniture (e.g., kitchen) and electrical installations, switches, light systems, and radiators were preinstalled. The completed modules left the factory and were transported to the construction site, JIT and JIS. The final assembly took place in a temporary, mobile, and heated assembly hall. Four assembly workers and one assembly foreman were needed per building. All production was performed indoors, either in the factory or in the mobile assembly hall. The factory was clean, and the components were handled by workers wearing gloves to protect the finished modules.

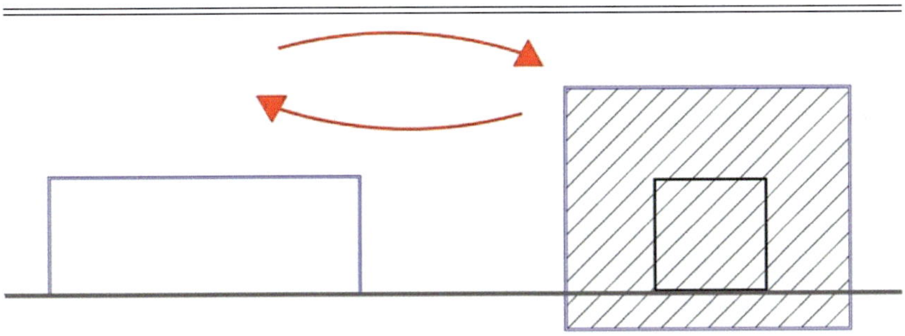

	System name	Company
16	NCC-Komplett	NCC

2.6.1 NCC Komplett

Company: NCC, Sweden
Category: Off- and on-site combined factory (both in a fixed location)

Analysis of Technology and Composition

Evolution Scheme From 2002, NCC Sweden began working on developing a concept for the automated production of multistorey residential buildings (Figure 2.326) and, in accordance with this concept, from 2005 to 2007 it ran a test project called NCC

Figure 2.323. On-site factory, exterior view. (Image: NCC)

Komplett. NCC Komplett (NCC Complete) is a way of building apartment blocks by an integration of off-site and on-site manufacturing. The system significantly reduced the construction time, enhanced quality, and reduced costs, while giving the architects design freedom for satisfying customer requirements. About 60 operators are used in the factory in shifts and in accordance with a rotating time schedule. The company's own concrete plant guaranteed the quality and steady JIT JIS delivery of the high-performance concrete elements that were required. The system was intended to have an annual capacity of at least 1000 apartments. Cycle times of operations in off-site prefabrication and on-site manufacturing could be fully synchronized because both were operated by the same company. The working environments in both off-site and on-site factories provided an SE and was clean (thus quality fostering) and ergonomically optimized (worker friendly and efficiency fostering). A speciality of the system was the deep integration and direct link between off-site factory and on-site factory.

Elevation and Ground Plan To protect the prefabricated and with finishing and furniture equipped panels from weather influence, to be independent of seasonal conditions, and to be able to conduct on-site assembly potentially in a 24/7-mode, the entire assembly process on the construction site took place in a hall-like on-site factory that encloses the whole building and some additional logistics space around it (Figure 2.323). The enclosed on-site environment provided an optimized, heated working environment. OMs were used to pick up the components from the trucks that enter the on-site factory and to position them. The OMs were the central element of the assembly process, lifting module after module into place. Five employees assembled three to five apartments per week. The assembly workers in the on-site factory wore gloves and shoe covers so as not to damage the finished surfaces of the prefabricated elements during their assembly on the site. Before element assembly, the basement or bottom slab was built in a conventional manner in the on-site factory. After assembly, the facade was finished by conventional painting. The on-site factory was heated to create an optimized work environment for the handling and positioning of the prefabricated concrete elements and for optimized final hardening of the prefabricated concrete elements.

Subsystems Off- and on-site combined manufacturing systems consisted of a set of integrated subsystems:

1. *Off-site component factory*. The off-site factory (Figures 2.327–2.332) prefabricated concrete elements that were equipped in the factory with all installations and finishing (cabling, windows, radiators, switches, kitchen and bath furniture/equipment).
2. *Component transportation system*. A large number of transport trucks delivered the components JIT and JIS, from the factory to the site.
3. *On-site factory structure*. A temporary on-site steel structure covers the whole work space, provides a weather-protected and structured environment and also a platform for the installation of OMs (Figure 2.325). The factory was self-supporting and rested on the ground. The factory structure also provided a weather-protected environment for excavation and basement work.
4. *On-site material handling, sorting, and processing yard*. An MHSPY was integrated into the on-site factory. Delivery trucks were able to drive directly into the factory and the material could be picked up directly from the trucks by the OM system.
5. *Overhead manipulator system*. In the interior of the factory large overhead rail-guided manipulators operated (Figure 2.324). They were mounted to the factory structure.
6. *Material flow management system*. NCC Komplett fully integrated demand oriented off-site manufacturing with the on-site factory. Components were delivered from the factory to the site JIT and JIS. All input materials came directly from material manufacturers. IT-based systems were used for fault-free project planning. Using state-of-the-art 3D software, architects created buildings with unique characteristics and designs. NCC's own concrete plant guaranteed the quality of the (for this construction method required) highly accurate and to a very high degree prefabricated concrete panels.

System Variations Due to the size limitations imposed by the on-site factory (SF structure), the maximum building height allowed was eight floors and the maximum length allowed was 60 metres.

Robot-Oriented Design On component level ROD was applied to modularize the building into as less high-level components as possible. Thus complex work tasks (including the equipping of the concrete components with cables, radiators, finishing, and furniture) were in majority shifted to the factory reducing work task variety and complexity on the construction site. With this design approach the basis for the deployment of a structured on site factory environment was laid. Furthermore, to allow the application of the on-site factory, on building level the design had to consider that the building can be assembled within the factory leading to separate condominium blocks instead of condominium rows. This design necessity also influenced the arrangement of the buildings on an urban level.

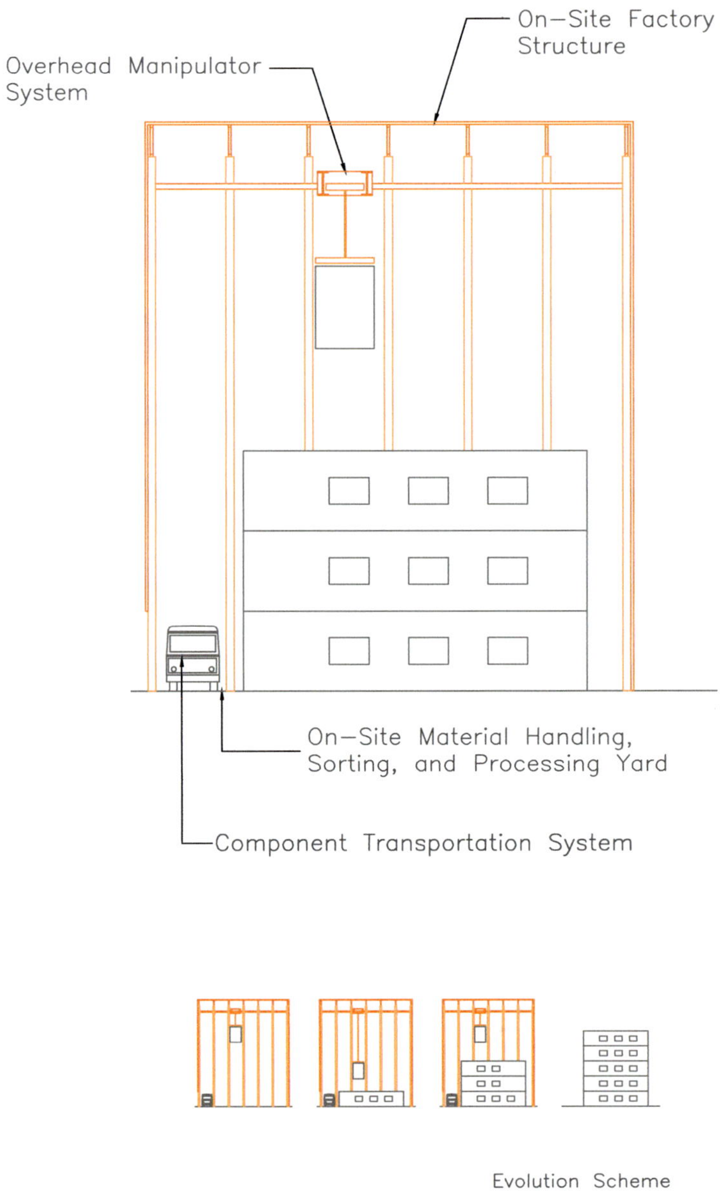

NCC Komplett, Company: NCC, Sweden

Figure 2.324. On-site factory, interior view. (Image: NCC)

Figure 2.325. On-site factory, exterior view. (Image: NCC)

Figure 2.326. Finished product. (Image: NCC)

Figure 2.327. On-site assembly of prefabricated high level concrete-based components. (Image: NCC)

Figure 2.328. Off-site factory. (Image: NCC)

Figure 2.329. Component prefabrication off-site. (Image: NCC)

Figure 2.330. Component prefabrication off-site. (Image: NCC)

Figure 2.331. Component prefabrication off-site. (Image: NCC)

Figure 2.332. Component prefabrication off-site. (Image: NCC)

2.7 Self-Supported Ground Factory (horizontally moving)

Systems in this category are used more for the assembly of horizontally oriented buildings than for vertically oriented high-rise buildings. Systems in this category are therefore characterized by a self-supporting GF, which completely encompasses the building. The factory moves on rails along the horizontal main direction of the building, subsequently allowing the completion of the building in a horizontal direction. Both systems assigned to this category are mechanized rather than automated systems that can be more considered as historical prototypes than as automated construction sites. However, they were considered for analysis in this chapter and for being integrated into the categorization systems, as the approach, especially for Europe (where buildings are in most cases horizontally oriented), is highly relevant. The author of this thesis sees a high potential to deploy such systems quickly and cost efficiently by utilizing the latest automation and robot technology.

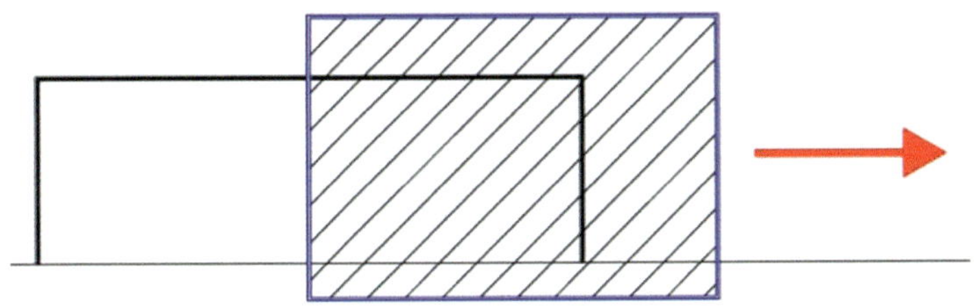

	System name	Company
17	Bauhelling	AHAG-Sommerfeld
18	Bauschiff	Neufert

2.7.1 *Bauhelling* Summerfield

Company: AHAG-Sommerfeld, Germany
Category: Self-supported GF (horizontally moving)

Analysis of Technology and Composition

Evolution Scheme, Elevation and Ground Plan Sommerfeld's *Bauhelling*, which provides the possibility of operating multiple cranes in parallel, was inspired by Harland & Wolff's shipbuilding yard, which constructed the titanic in 1913 (see also **Volume 1**). Adolf Sommerfeld introduced the *Bauhelling* system in the 1920s – initially inspired by shipyard construction (*Bauschiffe* in German). Before the construction process starts, a set of tracks is installed for the transportation of components and equipment to and on the site, and for moving the field factory frame structure on the site (Figure 2.333). The operation of the system starts with the installation of the field factory's frame structure, integrating the HDS and end-effectors. While one HDS places the formwork for the first floor of the building, the second half of the field factory's frame structure containing the VDS, the concrete mixer, and the concrete pouring machine are installed. Afterwards, the field factory frame structure moves forward and the end-effectors pour concrete into the installed formwork, while the

Figure 2.333. Set up of factory and logistics system on-site. (Image: Kress, 2011, p. 171/ Picture Documentation by AGAG-Sommerfeld, Schild)

hook-crane of the second half of the filed factory prepares the formwork in the other building section. When a whole floor is complete, the process begins again for the next floor level. The building is thus erected layer by layer, starting with the ground floor.

Subsystems The Sommerfeld *Bauhelling* system consists of a set of integrated subsystems:

1. *Guiding rails (tracks).* A set of tracks is installed on-site to enable the filed factory frame structure to move during the construction process back and forth, and to enable it to be transported to the next building in a row after the first has been completed (Figure 2.333).
2. *Field factory frame structure.* A frame structure (Figure 2.338) that can be moved on tracks and spans the whole building serves as a platform for the installation of overhead cranes (HDS) and vertical lifts (VDS). In addition, the frame structure can optionally be covered to provide a weatherproof work environment.
3. *Vertical delivery system.* A lifting system tower is attached to the factory's frame structure for bringing concrete and other materials to the HDS and the end-effectors (Figure 2.334).
4. *Horizontal delivery system.* The OMs (Figure 2.334) each span a section of the building and are able to to cover the whole workspace of the field factory frame structure. An OM has 3 DOFs (translation lengthwise, trolley hoist translation sideways, lower/lift of end link). Different end-effectors can be attached to the OM (e.g., crane hooks and concrete distributors).

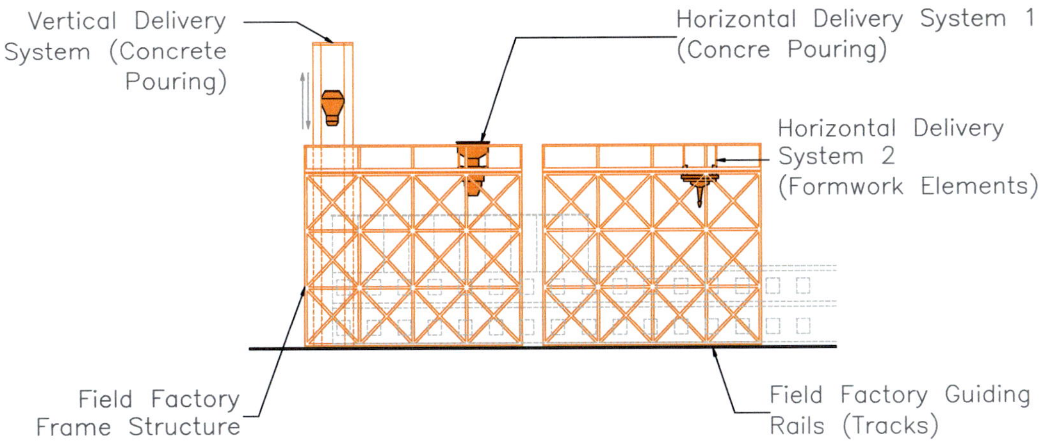

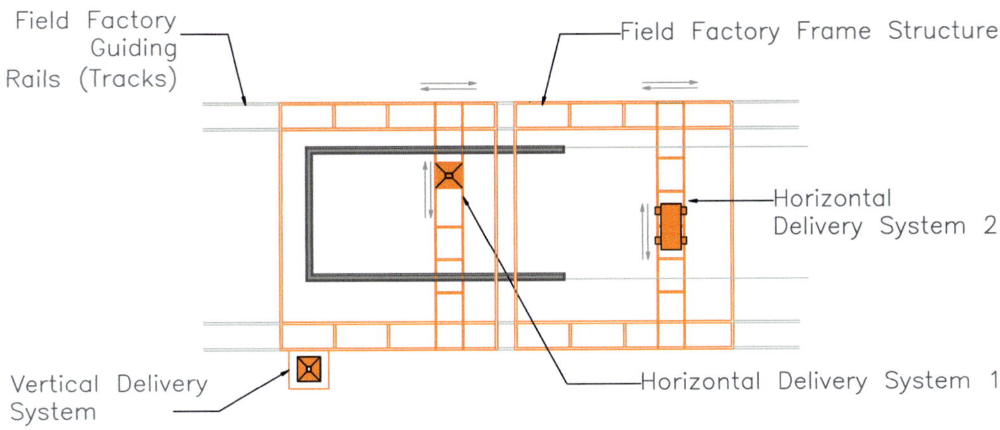

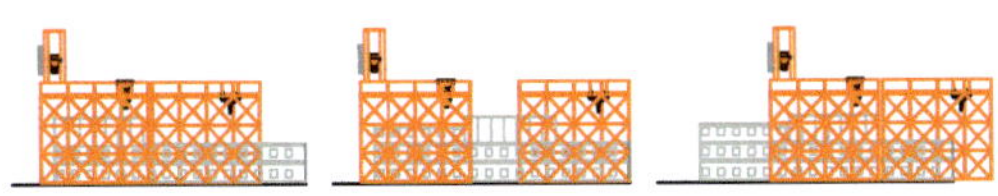

Bauhelling Summerfield, Company: AHAG-Sommerfeld, Germany

End-effectors The OMs can be equipped with following end-effectors:

1. A multipurpose hook end-effector is used for the transportation and placement of the formwork needed for the construction of the building.
2. A concrete pouring end-effector is used for placing the concrete delivered from the VDS over the OMs into the formwork.

System Variations The width of the factory frame structure can be varied. Furthermore, it is possible to extend the length of the factory frame structure (and to add, for example, additional OMs) or to increase its height for the production of taller buildings.

Figure 2.334. Operation of several filed factories in parallel. (Image: Kress, 2011: p. 186/ Städtebau 1928, No. 7, p. 162)

Robot-Oriented Design On a component level the application of the system was supported by designing simple and rectangular shaped floor layouts (Figures 2.335 and 2.336). Those layouts allowed the efficient use of the formwork kit system and guaranteed that the HDS could be used to manipulate and position the formwork system properly. On a building level, it was required that the buildings size does not exceed the workspace (and flexibility limitations) of the factory frame structure. On an urban level, it is advantageous to design the master plan in a way so that the on-site factory can be utilized to construct not only a building block but a whole row or array of buildings (Figure 2.337).

Figure 2.335. Finished product. (Image: Kress, 2011, p. 171/Picture Documentation by AGAG-Sommerfeld, Schild)

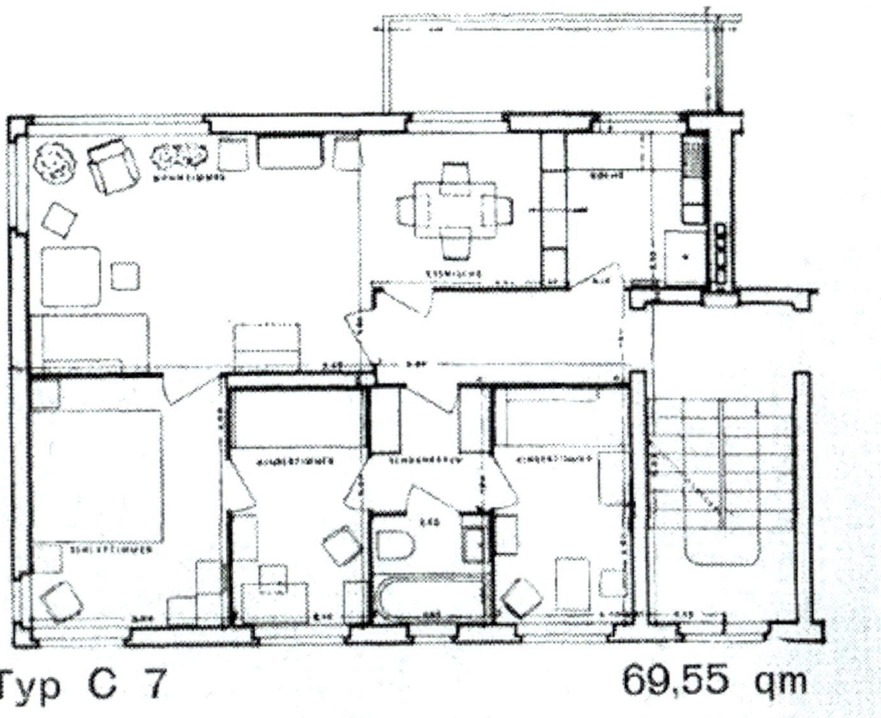

Figure 2.336. Simple and rectangular shaped floor layouts support the system application. (Image: Kress, 2011: p. 173/ Baugilde 1930, No. 16, p. 1462)

Figure 2.337. ROD on an urban level. (Image: Kress, 2011: p. 170/ Baugilde 1930, No. 16, p. 1460)

Figure 2.338. Field factory frame structure. (Image: Kress, 2011: p.168/ W. Block)

2.7.2 Bauschiff

Company: Neufert, Germany
Category: Self-supported GF (horizontally moving)

Analysis of Technology and Composition

The analysis of this system was done on basis of Neufert's original sketches, drawings, and plans laid down in by Neufert in his book *Bauordnungslehre* (Neufert, 1943).

Evolution Scheme, Elevation and Ground Plan Neufert's *Bauschiff* (literally translated, "building ship") system consists of a moving weather-protected factory structure, movable exterior and interior formwork segments, a material elevator (VDS), an HDS, as well as other formwork elements (e.g., foldable roof formwork). The factory structure rests on wheels and moves along tracks, producing linear concrete buildings. The building process starts with the preparation of the site. After the laying of the foundations, the factory structure is erected, followed by the assembly of the exterior and interior formwork subsystems. Concrete is transported to each floor via HDS and VDS. Once the hardening of the concrete is complete, the factory frame structure and the formwork subsystem are able to move along in the construction direction. After the concreting in the first section of the building, the work on each level follows slightly behind the work on the level below. While the lower formwork elements are in place and the concrete is being poured, the formwork on the next level is removed and the concrete wall is treated and finalized. At the same time, finishing work continues in the interior of the building that has been left behind. As the system moves along, it uncovers only the finished parts of the building. The formwork on each floor is able to move independently, starting from the ground floor and thereafter on the upper floors. Overhead tracks (part of the HDS) along this formwork frame allow the movement of concrete and other materials to the desired destinations. Work in the completed part of the building proceeds in parallel and the finishing work is done independently of the structural work for work parallelization purposes.

Subsystems Neufert's *Bauschiff* manufacturing of buildings on-site consists of a set of integrated subsystems:

1. *System wheels.* The site factory structure rests on wheels that move it along the building to be constructed. The wheel mechanism is guided by a temporarily erected ground-based rail system.
2. *Site factory structure.* The site factory structure provides a weather-protected environment on the construction site. The main building process takes place within this factory structure until the roof of the specific building section is completed.
3. *Vertical delivery system.* At the open end of the building, the VDS is used mainly to transport pre-filled concrete containers to the construction levels. On the ground level, concrete mixer trucks fill the concrete containers, which are then moved upwards via the VDS. The VDS delivers the containers with trolley hoists carrying them, and on the individual floors these trolleys can be driven directly over to the HDS allowing an uninterrupted flow of material.
4. *Formwork segments.* Used on each floor; move linearly and in parallel during the building process.
5. *Horizontal delivery system.* An overhead rail system allows trolley hoists to carry the concrete containers on individual floors to the desired position.

End-effectors

1. *Concrete containers.* The concrete tanks are are moved on the site through HDS and VDS.
2. *Scissor-gripper.* Scissor grippers are used to pick and transfer through HDS and VDS.

System Variations The width of the factory frame structure can be varied. Furthermore, it is possible to extend the length of the factory frame structure (and the formwork subsystem) or to increase its height for the production of taller buildings.

Robot-Oriented Design The window openings are integrated in the formwork moved forward in this system along the direction of construction, demanding a relatively systematic and monotonous design of the façade of the building. Furthermore, from an urban perspective the system could be efficiently used to construct long, horizontally oriented rows of buildings or master plans with interrupted/uninterrupted long building blocks, orthogonally arranged along a horizontal axis.

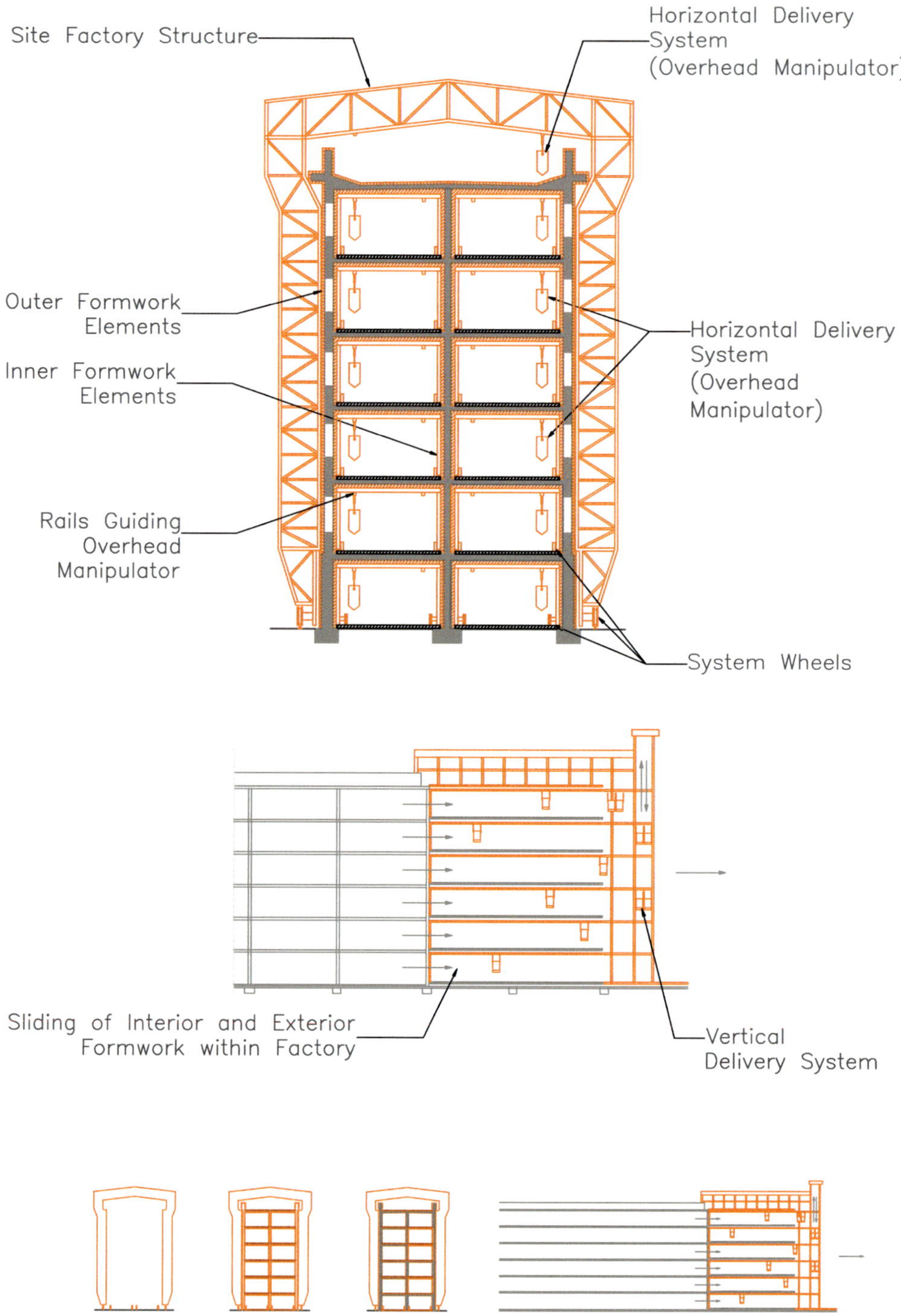

Bauschiff, Company: Neufert, Germany

2.8 Sky Factory (moving upwards) for Simple Tower Manufacturing

Systems in this category manufacture simple (noninhabitable) tower structures with a highly repetitive structure by using a factory that moves from bottom to top. In Japan in particular, there is a high need for such towers, which are used as advertisement towers, broadcast towers or sightseeing towers. Both systems in this category are simplified and altered versions of systems used for automated building construction. Systems in this category can, in contrast to other automated construction sites, be identified as line construction sites, as assembly is achieved by very closely following a vertical axis and because a high amount of like or similar components are assembled by a central manipulation system.

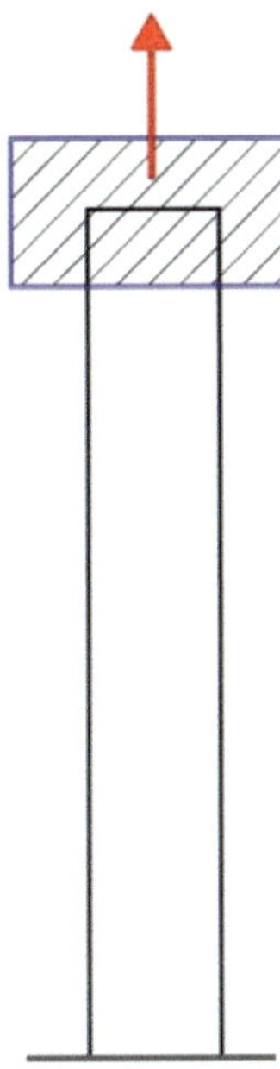

	System name	Company
19	TS-Up	Taisei
20	Tower SMART	Shimizu

2.8.1 TS-Up

Company: Taisei Corporation, Japan
Category: SF (moving upwards) for simple tower manufacturing

Analysis of Technology and Composition

Evolution Scheme Taisei uses the TS-Up (Figure 2.339) to manufacture (noninhabitable) towers for broadcasting and advertisement. The TS-Up makes the manufacturing of such towers weather immune, faster and prevents disturbance to the environment by construction. The system can be set up to construct a tower on top of an existing building, on the ground to construct a free standing tower, or a tower that is

Figure 2.339. Sky factory.

partly integrated as a servicing shaft into a building. A tower crane beside the tower positions the vertical structural main columns. Those are temporarily fixed into position and used as the structure along which a self-climbing two level OSF is lifted. The SF provides working platforms on the interior of the tower (which, during the positioning process, also function as fixtures for the guiding of prefabricated bracing elements into place) and working platforms on the outside for fixing and welding. The factory works along the vertical axis bidirectionally, as the factory is capable not only of climbing up the outside of the tower as it rises, but after preliminary completion can also climb down the structure. The step-by-step procedure of climbing down is used to weld joints to bracing elements at each level, and also to coat welded areas with alloy and paint. In addition, during the operation of moving down, the upper platform of the SF can be used to assemble platforms to the tower.

Elevation The elevation shows that work during the upwards rising process is done on three levels (the *n*th level, $n - 1$, $n - 2$). The *n*th floor is only at the very bottom part covered by the SF and the structural main columns are positioned on this level. On the $n - 1$ level, the factory provides working platforms on the outside and inside of the building that function as fixtures during the guidance into place of prefabricated bracing elements during their insertion. On level $n - 2$, the factory provides working platforms on the outside and inside of the building for fixation welding operations.

Also on level $n - 2$, the climbing mechanism is integrated into the factory and grips the tower structure. During the lifting of the factory from the top of the tower, the two working levels of the factory, and the multiple platforms, provide a safe working space for further fixation and coating.

Ground Plan The tower crane, positioned besides the tower to be erected, has a large working space, and allows the storage or delivery space to be located some distance from the tower (in contrast to Tower-SMART, for example, where factory intake port requires material to be delivered directly to the ground of the tower). This simplifies the erection of towers in a confined space or on top of buildings.

Subsystems The TS-Up manufacturing the building on-site consists of many integrated subsystems:

1. *Sky factory outer stages*. The SF (Figures 2.339 and 2.340) outer stages cover two segment levels that provide working platforms and integrate the climbing mechanism that raises/lowers it.
2. *Sky factory inner stages*. The inner part of the SF (within the tower) consists of 24 vertical stages (every second level, with 6 working platforms per stage). They are lifted and lowered by the tower crane, which can reach the inner part of the SF from the top.
3. *Climbing mechanism*. Four hydraulically actuated inchworm climbing mechanisms are positioned in the four corners of the tower. They climb up the structural main columns, which have been positioned, erected, and fixed beforehand.
4. *Vertical/horizontal delivery system*. A tower crane is extended vertically during the erection of the tower. It is positioned right beside the tower and is mounted to the tower on every second level. The tower crane consists of segments and in order to be extended vertically new segments are added. The tower crane has a large working space and allows for the storage or delivery space to be located some distance from the tower.

End-effectors The V/HDS uses simple (conventional crane-like) multipurpose end-effectors. The fixture guiding components into place in the interior of the tower can be seen as an end-effector as well.

System Variations The TS-Up was developed on the basis of concepts and technologies of Teisei's T-Up. The system can be adjusted to towers that are similar in terms of the bearing structure, but that have, for example, different heights or distances between the vertical main columns. The system is currently only capable of processing steel components.

Robot-Oriented Design: The tower is modularized into a kit that is adjusted to the needs and operational sequences of the TS-Up. The kit consists of:

- Vertical main structure column (height: exactly one SF level) with elements already fixed
- Horizontal beam elements
- Rectangular bracing elements
- Staircase elements and prefabricated platform elements

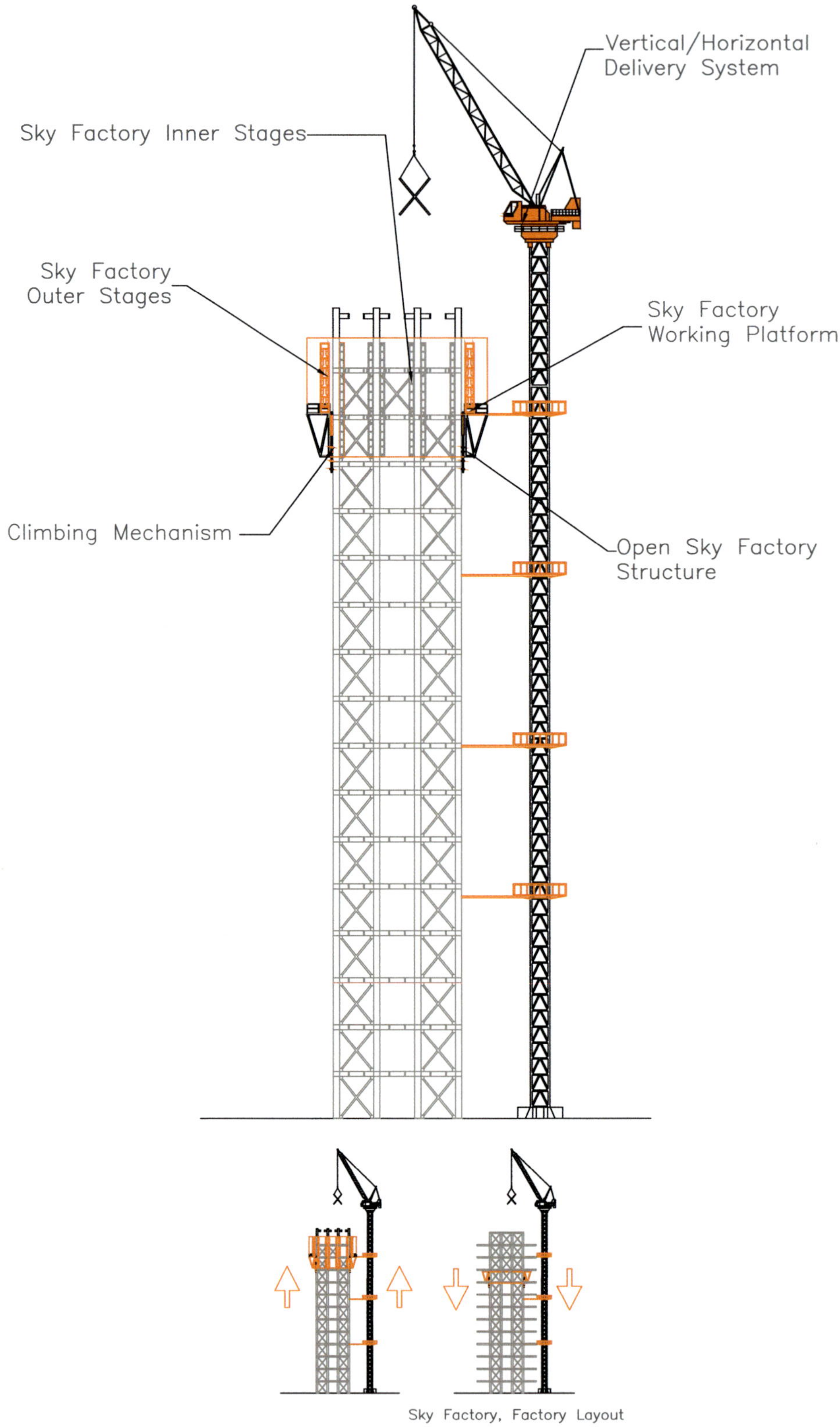

TS-Up, Company: Taisei Corporation, Japan

Figure 2.340. Sky factory, detailed view.

Figure 2.341. Sky factory. (Image: Shimizu)

2.8.2 Tower-SMART

Company: Joint venture of Shimizu Corporation and NTT Facility Corporation, Ltd., Japan

Category: SF (moving upwards) for simple tower manufacture

Analysis of Technology and Composition

Evolution Scheme Shimizu developed the Tower-SMART (Figure 2.341) jointly with NTT Facility Corporation, Ltd. To construct broadcast towers all over Japan (small and medium sized towers, to support, for example, mobile phone networks) for NTT DoCoMo. The Tower-SMART makes the assembly of such towers weather immune and shortens the time necessary for construction by 20%. For the set-up of the Tower-SMART, the lower part of the tower (i.e. the first 5–10 metres) has to be built conventionally so that a stable and stiff base for the attachment of the CS is provided (see Figure 2.342 for details about the evolution scheme). The CS is able to grip these conventionally constructed and fixed columns from the outside and pushes up the factory. As the CS climbs up the outside of the tower (and does not sit on top of the columns), after the tower has been completed, the climbing mechanism can simply climb down the tower again and be disassembled at the ground level.

Elevation A trolley manipulator can travel along two gantry cranes. The trolley manipulator travels to the smaller gantry crane to lift up steel components from the ground, and then travels over to the larger gantry crane, which spans the whole floor area of the tower. The steel components are high-level components, reducing the amount and complexity of assembly operations in the SF.

Ground Plan The ground plan shows the interplay of the two gantry systems. The smaller one operates over the intake port and guides the trolley manipulator during component lifting. The larger of the two guides the trolley manipulator during the component positioning process. The SF within the cover provides reference points for the installation of the components as well as working platforms for the workers attaching welding robots and/or welding themselves.

Subsystems The Tower-SMART manufacturing broadcast towers on-site consists of a set of integrated subsystems:

1. *Factory roof structure.* The SF (Figure 2.341) completely covers the place where components are assembled on top of each other. On one of the four sides, the SF cantilevers over the floor area of the tower, generating a covered intake port through which material can be delivered. Rail-guided overhead systems can be fixed to the roof. The SF also provides working platforms on which workers can support the positioning, alignment, and fixation process.
2. *Sky factory interior working platform.* The SF provides working platforms which allow column components to be conveniently positioned, aligned, and fixed (Figure 2.343).
3. *Sky factory column positioning template.* The SF functions as a template for positioning of column components by providing reference structures in its interior (Figures 2.344–2.346).
4. *Climbing system.* An inchworm-like and hydraulically actuated CS sits at the lower end of the factory. Upon the CS rest four round steel columns, which are the main structural elements of the SF. These four main columns are enclosed by a steel structure that is the basis for the cover, and it also provides the basis for the factory roof, factory floor, and working platforms to be mounted. The CS pushes the factory upwards along previously positioned and fixed column components (Figure 2.347).
5. *Vertical delivery system.* On the factory roof, over the intake port, a gantry crane operates on which the trolley manipulator can travel and lift up components from the ground (Figure 2.347). The trolley manipulator and gantry crane together provide 3 DOFs (translation sideways by gantry crane, travelling of trolley manipulator along the gantry rail, lifting/lowering of end link). The end-effector adds additional DOFs to the system.
6. *Horizontal delivery system.* On the factory roof, above the floor area, a gantry crane operates on which the trolley manipulator can travel and position components. The trolley manipulator and gantry crane together provide 3 DOFs

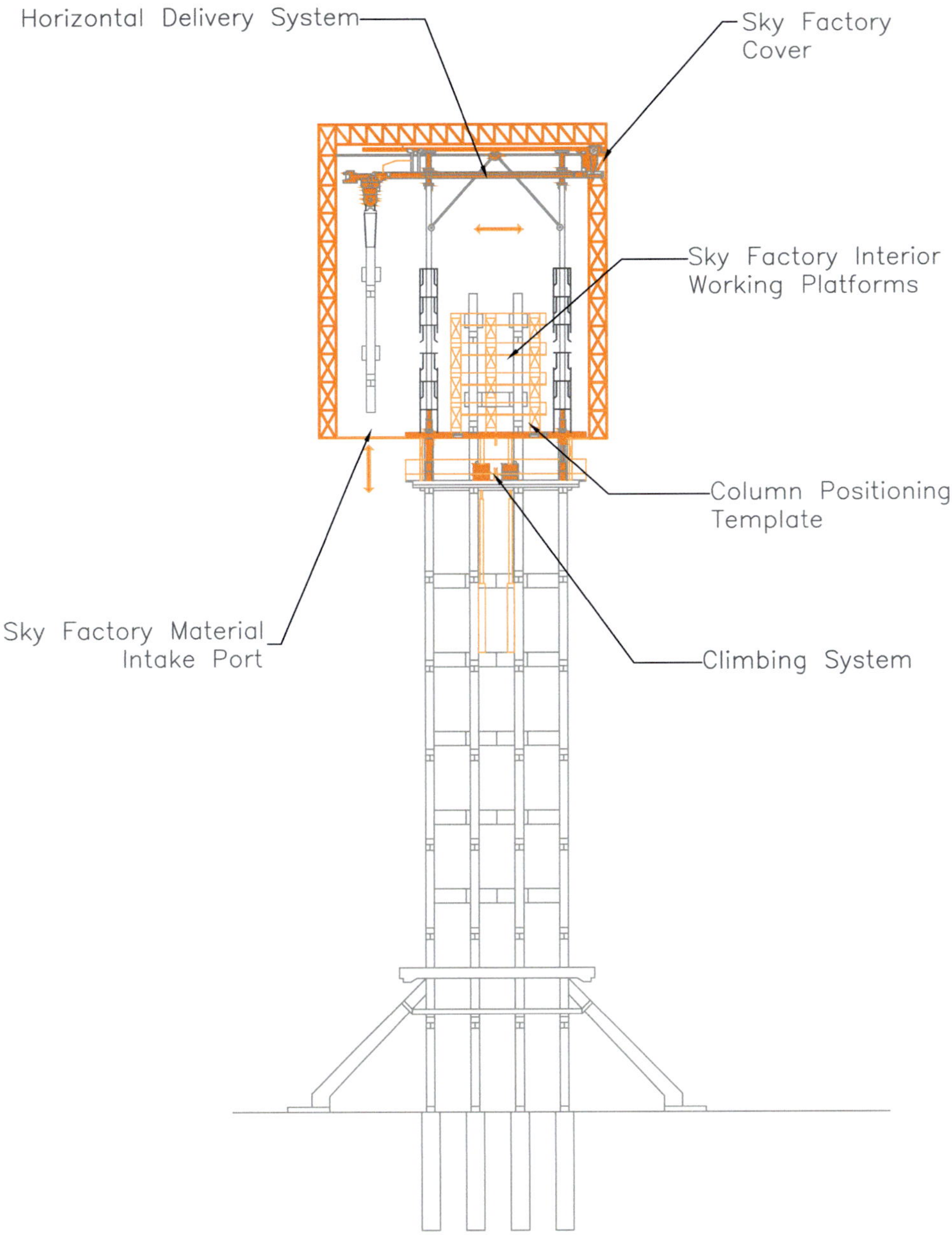

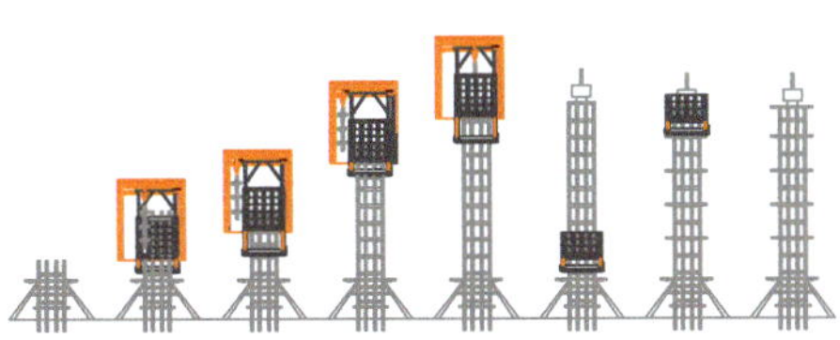

Evolution Scheme

Tower-SMART, Company: Joint venture of Shimizu Corporation and NTT Facility Corporation, Ltd., Japan (translation sideways by gantry crane, travelling of trolley manipulator along the gantry rail, lifting/lowering of end link). The end-effector adds additional DOFs to the system for orientation the component.

7. *Welding robot system (optional)*. Welding robots can be used to fix the steel components to each other.

End-effectors Because ROD reduces the complexity of the installation of high-level vertically orientated column elements, the need for various end-effectors is erased. The trolley hoist has an end-effector that holds the column during the lifting, positioning, and adjustment processes by four ropes; each rope is attached to one of the four sides of the column, where small steel plates welded to the component provide gripping points.

System Variations The system itself is based on concepts and technologies developed before for the SMART. Whereas in the SMART multiple trolley manipulators work in parallel, in the Tower-SMART only one trolley manipulator is used. In addition, the design of the steel components according to ROD, and the resulting reduction in complexity in the SF goes a step further than in the SMART. From this perspective the Tower-SMART represents a simpler and less complex system.

Robot-Oriented Design As the Tower-SMART is a Joint venture of the construction company (Shimizu) and the client (NTT DoCoMo), the design of the towers on a component, building and urban level can be fully adjusted to the capability of the Tower-SMART. The columns to be positioned are high-level components, which reduces complexity in the SF, and which are designed especially for installation by the system. The columns already have multiple beam elements fixed to them, so that the positioning of additional horizontal/beam elements is not necessary in the SF. The OM only needs to install the vertically oriented high-level components. The components at the bottom have steel clamps attached on all four sides that support the adjusting process. On a building or urban level, the applicability and efficiency of the Tower-SMART can be supported by synchronizing the location of the tower and the allocation of the Tower-SMART's SF intake port and its accessibility from below.

Figure 2.342. Evolution scheme photographs. (Image: Shimizu)

Figure 2.343. Column positioning by an overhead manipulator. (Image: Shimizu)

Figure 2.344. Sky factory interior. (Image: Shimizu)

Figure 2.345. Column positioning by an overhead manipulator. (Image: Shimizu)

Figure 2.346. Column positioning by an overhead manipulator. (Image: Shimizu)

Figure 2.347. Overhead manipulator lifting a column component up into the sky factory via the intake port. (Image: Shimizu)

2.9 Centralized Sky Factory (moving upwards) in Combination with Conventional Construction

Systems in this category are characterized by the application of centralized SFs, which manufacture strategically important parts of the building. Around the centralized SF – in parallel – other parts of the building are constructed conventionally. However, the central SF – over the vertical logistic system – plays a major part in terms of material supply to the individual floors. Obayashi (see below) analysed the efficiency of the location of the SF (identifying which parts were covered by conventional construction and which by SF) and proposed and deployed the system in various configurations. The system and its subsystems were based on the ABCS system allocated to the category "SF (moving upwards) supported by Building" (see Section 2.1). The Hybrid-ABCS can thus also be considered as a variation of ABCS.

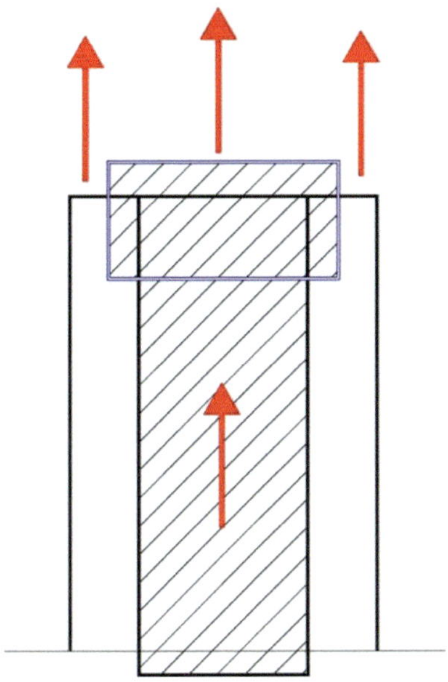

	System name	Company
21	Hybrid-ABCS	Obayashi
22	Hybrid-SMART	Shimizu

2.9.1 Hybrid Automated Building Construction System (Hybrid-ABCS)

Company: Obayashi Corporation, Japan
Category: Centralized SF (moving upwards) in combination with conventional construction

Analysis of Technology and Composition

Evolution Scheme The Hybrid-ABCS (Figure 2.348) is applied to buildings with a large floor area. To use the capacity of the automated OMs optimally, the SF is erected only over a strategically important area of the building (e.g., the core area of a building containing stair cases, elevators and major parts of the building technology). The floor area surrounding this strategically important area is erected in parallel, not by use of the OMs but by the use of rail-guided (optionally robotic) jib cranes, human labour, and logistics support from the SF. As the assembly of structural components outside of the strategically important area over which the SF is erected is performed by using more human labour than in the main area, the system is called "hybrid" (automated construction and human-based construction in parallel). The combination of both methods in one system turned out to be more efficient (*Efficiency Analysis* that follows) than the ABCS building which relied solely on automation. Consequently, parallelization of work in the core area and the surrounding area is indeed possible.

Elevation The elevation shows that an ABCS SF with three OMs side by side is usually installed over the central area. A VDS (automatic lift) feeds the SF with material.

Figure 2.348. View of a sky factory. (Image: Ikeda & Harada, 2006)

The outer area is spanned by two rail-guided jib cranes, which are mounted to the top of the SF. As the outer area is not constructed automatically by the OMs, side covers enclosing the factory are not necessary, reducing factory set-up and factory disassembly time.

Ground Plan The ground plan shows that the SF spans the core area of the building, as well as a side part. The allocation of the factory thus allows for the more complex core assembly part to be done or supported by the OMs, for an automatic lift to be installed and connected to the SF, and for the not so complex office areas surrounding the core to be erected in parallel by the more human-based methods.

Subsystems The Hybrid-ABCS manufacturing the building on-site consists of many integrated subsystems:

1. *Sky factory structure*. To use the capacity of the automated OMs optimally, the SF is erected only over a strategically important area of the building. The SF provides a platform for the attachment of the OMs, the rail-guided jib cranes and the CS (Figure 2.353).

2. *Climbing system*. The factory rests and climbs on top of the building's steel columns by hydraulically powered steel stilts that bear the weight of the roof, pushing it upwards floor by floor. Stilts are positioned over each column; statically, only every second of the steel stilts needs to bear the weight of the factory, and it is thus possible for every second of the stilts to be lifted to enable the manipulator to position a new column under it. The CS is not used as a physically active fixture allowing the self-adjusting of the component (like with FACES), but it at least guides the installation process by providing reference points for the AAMS.
3. *Vertical delivery system*. An automatic lift delivers components (steel columns, steel beams, concrete floor slabs etc.) to the SF and to the OMs. The automatic lift is also used to supply surrounding area, via the SF, where the labour-based method is applied.
4. *Horizontal delivery system (overhead manipulators)*. Three gantry-type OMs, 4 DOFs (translation lengthwise, translation sideways, rotation, and lift/lower end link) with exchangeable component adjusted end-effectors. Manipulators are rail-guided; rails are mounted to the SF's roof structure.
5. *Vertical/horizontal delivery system (jib cranes)*. Two jib cranes are operated lengthwise along rails on top of the SF. The jib cranes can be operated via remote control or automatically.
6. *Welding robot system for columns*. The welding robot system for columns is a mobile STCR that has been integrated into the (integrated) site operation. The mobile system is moved over the pre-completed floors to the column to be welded, where it is positioned by and aligned with the column by the manipulator. The robot embraces the column and welds it simultaneously from two sides at once, so that any distortion of the steel column by the welding process is avoided. Several welding robot systems can be used to weld columns in parallel.
7. *Welding robot system for beams*. The welding robot system for beams is positioned by the manipulator on top of the preliminary fixed beams to weld/attach the beam components to the column components. The welding system consists of two frames used for the attachment to the beams and four welding robots with one welding head each mounted on top of these frames.
8. *Mobile (horizontally/floorwise operating) logistics robots*. On the floors within the SF already covered with ceiling elements, and that are thus no longer accessible to the OMs, mobile logistics robots are used for horizontal material delivery (mainly for the delivery of material for interior finishing). The logistics robots are integrated with the RTMMS and can communicate with the automatic vertical lifts to achieve JIT and JIS material delivery.
9. *Alignment and accuracy measurement system*. Sets of optical sensor systems (stereo vision systems, 1D and 2D laser scanning systems) are used to support the alignment and accuracy measurement process.
10. *Real-time monitoring and management system*. Data from the sensor systems, as well as from the servomotors/encoders of the vertical HDS, along with information obtained from cameras monitoring all activities (including human activities) in the factory, are used to create a real-time representation of equipment activity and of the construction progress. A barcode system allows representation and optimization of the material flow and allows equipment (e.g., the automatic lift)

to identify the component being or to be processed. Real-time monitoring and management is done in a fully computerized on-site control centre (see also *Efficiency Analysis* that follows).

End-effectors The manipulators (HDS) can be equipped with various end-effectors to suit the pickup, positioning, and adjustment of various types of components:

1. End-effector for picking up, positioning, and adjusting columns
2. End-effector for the horizontal delivery of multiple beams or steel profiles
3. End-effector for the positioning of beam/column components
4. End-effector for picking up and positioning concrete floor slabs
5. End-effector for delivery and positioning of the welding robot system for beams

System Variations The strategically important area of the building can be varied in location as well as in size. The SF is itself modular and is a physical and informational platform for the installation of subsystems in various configurations.

Robot-Oriented Design On a component level, the sizes, dimensions, and weights of components are designed so that they can be manoeuvred by the OM system. On a building level, the allocation of the SF in a dedicated and strategically important area is taken into consideration during planning. On an urban level, the location of the automatic lift, its accessibility to delivery systems such as trucks, and so forth on the ground level, and its relation to the SF can all be taken into account when deciding the orientation of the building in the environment.

Efficiency Analysis

The Hybrid-ABCS combines ABCS and conventional construction with the aim of combining the advantages of both methods and to achieve higher efficiency. The Hybrid-ABCS uses a central part of the building for the erection of the ABCS, and conventional labour-based assembly is used for the erection of the rest of the building. The central ABCS is also used to supply the conventionally constructed part with material. If ABCS was used for the whole building, the necessary long travelling distance of the automated – and not particularly fast – ABCS OMs would slow down construction (the building in Project 5 was very broad concerning its ground plan). By using both ABCS and conventional methods in parallel, both can operate simultaneously and nevertheless create synergies (e.g., logistics).

Rough Project Schedule For Obayashi, the Hybrid-ABCS represents an extension of the capability of ABCS. Its application in Project 5 (see Figure 2.351) was the beginning of a development in which Obayashi started to use ABCS elements along with conventional construction. After successful completion of Project 5, Obayashi even went one step further and started to use individual ABCS subsystems (e.g., only an automatic lift) in projects where predominantly conventional construction was applied (e.g., Tokyo Skytree, see Section 2.10.1). Today, Obayashi does not apply the full ABCS any more but rather its individual subsystems. It is remarkable that by applying the Hybrid-ABCS for the first time, Obayashi was able to completely construct a building with an average size of 3650 m^2 per floor, and 22 floors within six months (actual on-site construction time).

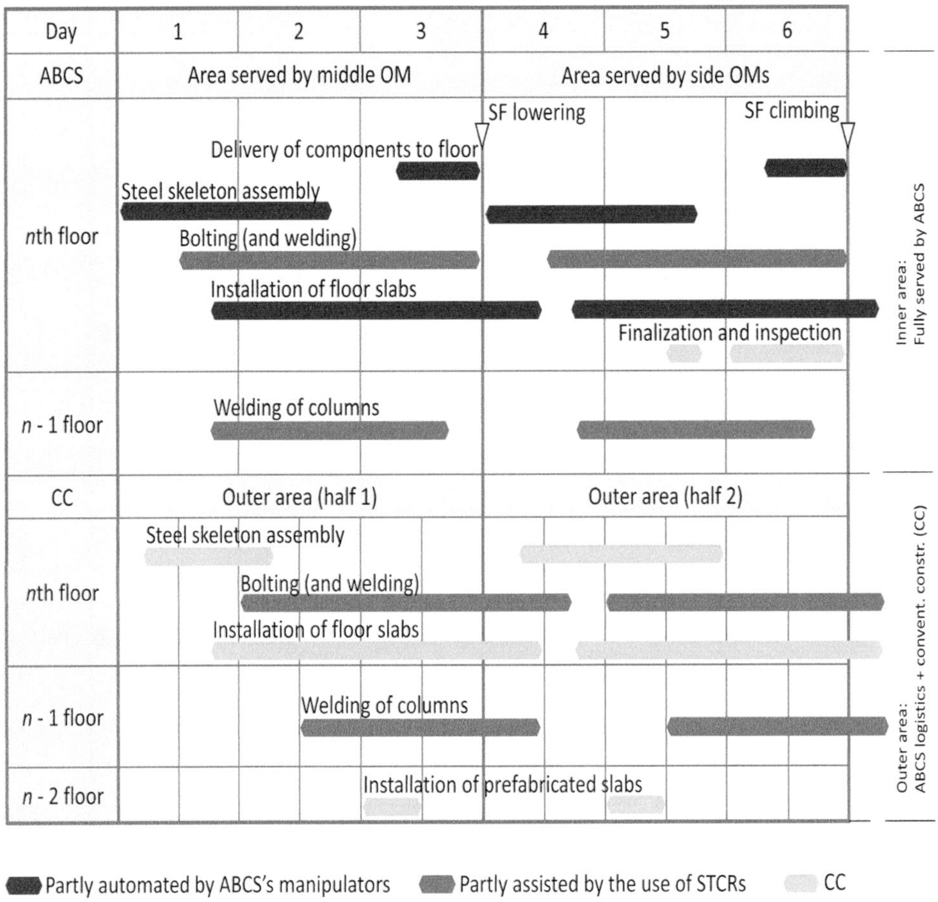

Figure 2.349. Schedule for standard floor erection cycle (6-day process) (adopted from Ikeda & Harada, 2006).

Floor Erection Cycle During the application of the Hybrid-ABCS, Obayashi experimented with both 5-day and 6-day cycles for completion of an average floor. As the 6-day cycle (Figure 2.349) was the norm, the experiments showed that a 5-day schedule was possible (Figure 2.350). The schedules showed that the application of ABCS and conventional construction methods (CCMs) in clearly defined parts of the building allows for parallelization of work.

Productivity The Hybrid-ABCS (Project 5) was able to reduce the requirement of human labour (man-hours/m^2) by more than 10%, compared to ABCS (Figure 2.351). Compared to conventional construction, ABCS had already reduced the requirement of human labour by about 40% (see ABCS, Section 2.1.1).

Learning Effects Obayashi recorded the time the workers needed to complete the welding operation on each floor (Ikeda & Harada, 2006; Figure 2.352). The number of welding operations on each floor was basically the same. The learning effect within the project was enormous, and the time needed to complete the welding tasks was brought down from 120% (4th floor) to less than 40% (20th floor).

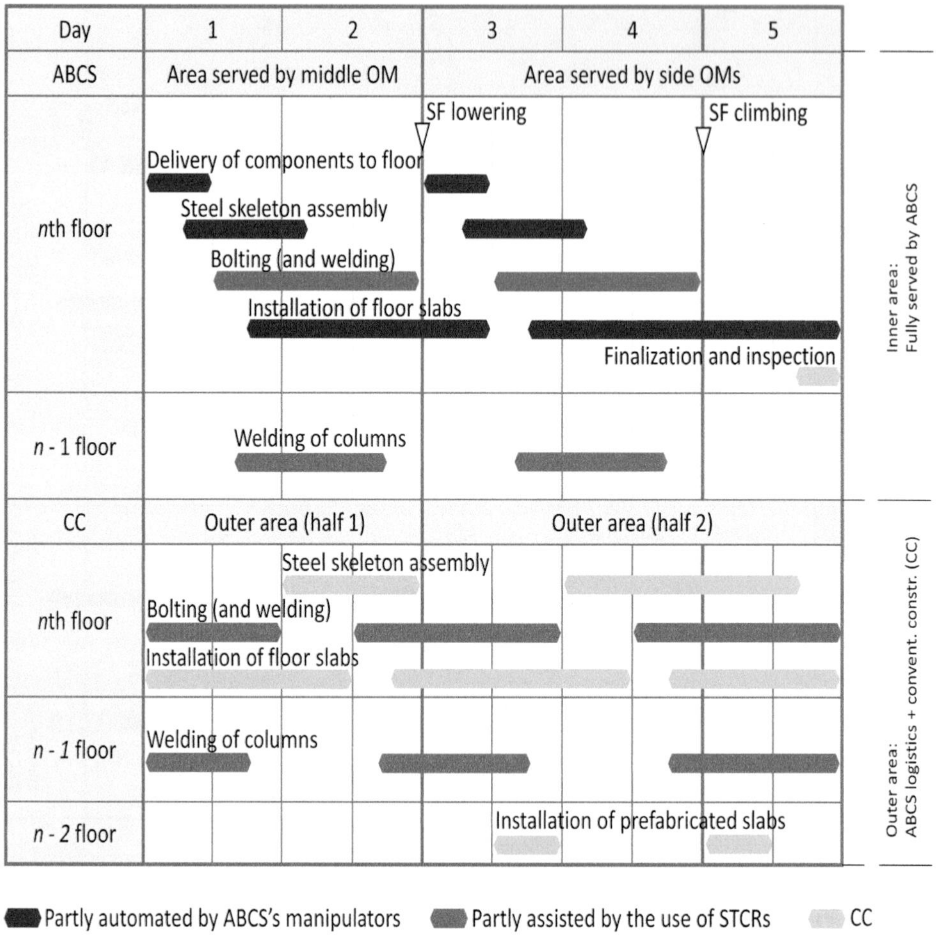

Figure 2.350. Schedule for shortened floor erection cycle (5-day process) (adopted from Ikeda & Harada, 2006).

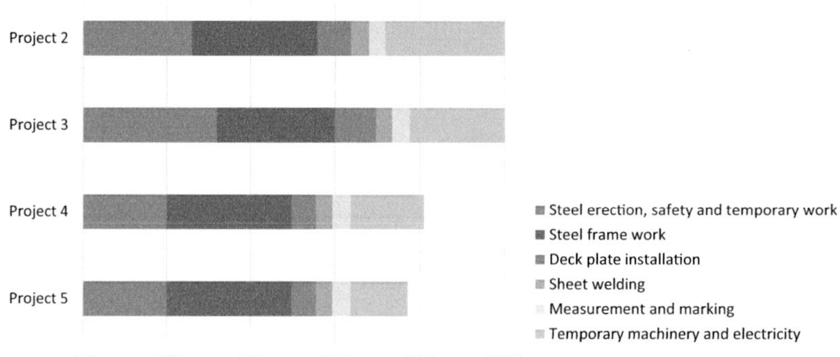

Figure 2.351. Comparison of labour requirement per floor area for different projects (Projects 1-4: subsequent ABCS applications, Project 5: Hybrid ABCS; see also Section 2.1.1) and work tasks (adopted from Ikeda & Harada, 2006).

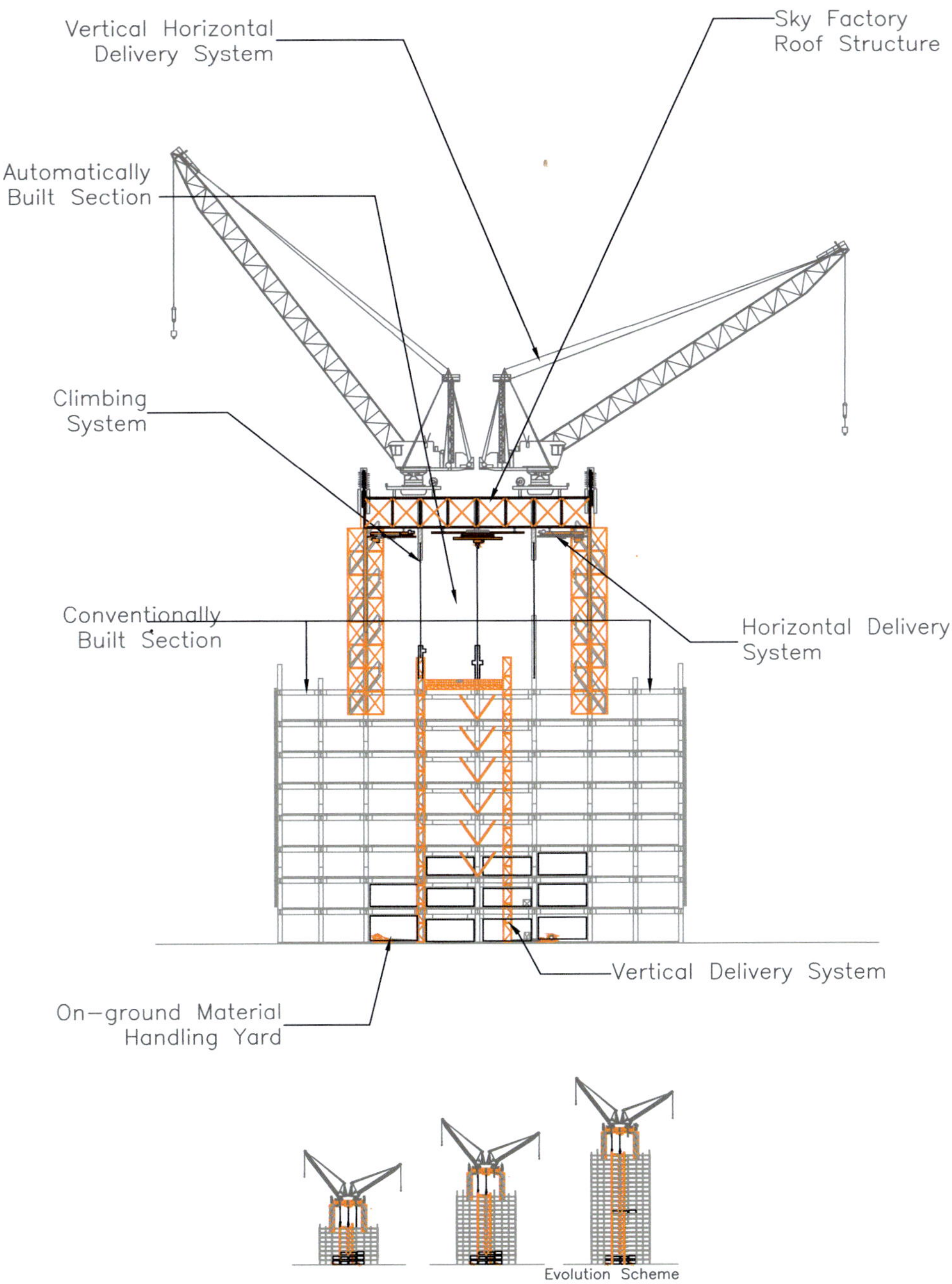

Hybrid Automated Building Construction System (Hybrid-ABCS), Company: Obayashi Corporation, Japan

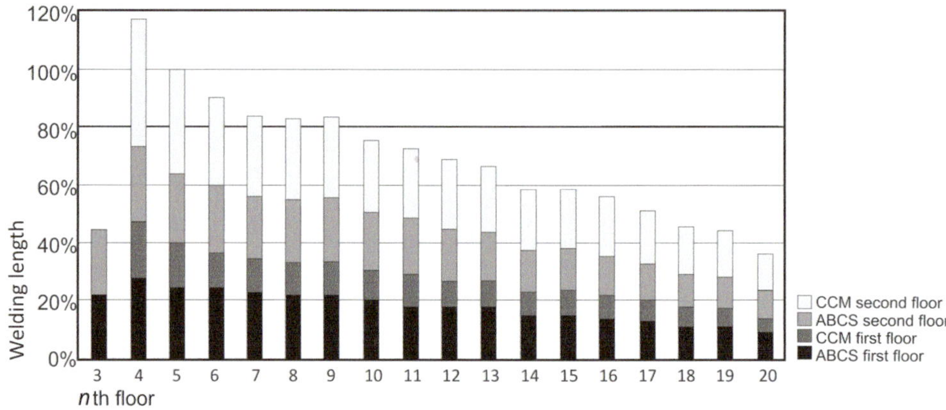

Figure 2.352. Comparison of time needed to and learning effects regarding welding tasks in the two operational floors (*n*th floors) between conventional construction and ABCS application; the 5th floor served as 100% example (adopted from Ikeda & Harada, 2006).

Figure 2.353. Interior view of a sky factory over the core. (Image: Ikeda & Harada, 2006)

2.9.2 Hybrid Shimizu Manufacturing System by Advanced Robot Technology (Hybrid-SMART)

Company: Shimizu Corporation, Japan
Category: Centralized SF (moving upwards) in combination with conventional construction

Analysis of Technology and Composition

Evolution Scheme The Hybrid-SMART (Figure 2.354) is applied to buildings with a large floor area. To use the capacity of the automated OMs optimally, the SF is erected only over a strategically important area of the building (e.g. the core area of a building containing staircases, elevators and major parts of the building technology). The floor area surrounding this strategically important area is erected in parallel, not by use of the OMs but by the use of rail-guided (optionally robotic) jib cranes, human labour and logistics support from the SF.

Elevation As in the SMART, the steel stilts that support the SF do not rest on top of the columns but between them – thus the climbing process interferes less with component assembly process. Within the Hybrid-SMART work is done on three to four levels in parallel. Components are positioned and aligned on the uppermost floor (the *n*th floor). Activities such as component fixation, interior finishing,

Figure 2.354. Sky factory, exterior view.

and facade work (over platforms provided by the SF) are done on the floors below.

Ground Plan The ground plan shows that the SF spans the core area of the building, as well as a side part. The allocation of the factory thus allows for the more complex core assembly part to be done or supported by the OMs, for an automatic lift to be installed and connected to the SF, and for the not so complex office areas surrounding the core to be erected in parallel by the more human-based methods.

Subsystems The SMART system manufacturing the building on-site by way of an SF consists of a multitude of integrated subsystems. The trolley hoist manipulators freely travel throughout the system by automatic vertical lifts and automatic overhead cranes. They thus integrate vertical and HDSs and take away the necessity of handing over components between them, thus eliminating waste processes (consider the waste that occurs in other systems as a result of the handover between vertical and HDSs, for example, the resulting idle time of the component and the time necessary to accomplish handover; in addition, consider the possibility of damage to components during handover).

1. *Sky factory structure*. To use the capacity of the automated OMs optimally, the SF (Figures 2.355–2.363) is erected only over a strategically important area of the building. The SF completely covers the automated construction area. The SF provides a platform for the attachment of the OMs, the rail-guided jib cranes (that serve the not-automated construction area) and the CS.
2. *Climbing system*. Four lifting stilts are positioned inside the factory to lift the whole factory. In contrast to many other systems that use SFs, the hydraulic lifting mechanism is not located in the roof structure itself but on the bottom of the stilts. Inchworm-like, the mechanism moves from floor to floor, pushing up the stilts and thus the factory.
3. *Trolley hoist overhead manipulator system (optional)*. In the Hybrid-SMART, multiple trolley hoist OMs are operated in parallel. Each trolley manipulator can travel between material handling systems on the ground and the SF's automatic overhead gantry crane system, thus integrating pick-up, vertical delivery, and positioning processes (see also SMART, Section 2.17).
4. *Sky factory horizontal delivery system (optional)*. Multiple mobile overhead cranes travel the building in a fixed lengthwise direction. However, the trolley hoist manipulators can travel between the cranes (see also SMART, Section 2.17). Thus, the trolley hoist manipulators can travel to any point on the *xy*-axis. Multiple overhead cranes can build a line (transport channel) along which the trolleys can travel.
5. *Gantry-type overhead manipulators (optional)*. Instead of the HDS simpler gantry-type OMs can be installed within the factory to perform component positioning operations (Figure 2.358).
6. *Vertical delivery system (lift of trolleys/manipulators)*. The trolley hoists pick up components from the ground floor and travel with the component via the automatic vertical lift into the factory (see also SMART, Section 2.17). In case the SF

is equipped with the simplified gantry-type OMs the automatic vertical lift only delivers components.

7. *Vertical/horizontal delivery system (jib cranes)*. Two jib cranes are operated lengthwise along rails on top of the SF. The jib cranes can be operated via remote control or automatically.
8. *Welding robot system*. Compared to other welding robot systems, the Shimzu welding robot system used within the SMART as well as the Hybrid-SMART is compact and lightweight. It consists of a ring which encircles the column and a single welding robot guided by this ring. Many of these systems can be used in parallel. Each welding robot system determines its welding path by a laser sensor and automatically welds the steel columns together.
9. *Real-time monitoring and management system*. Each construction element is tagged with a barcode. This code is scanned and the individual element is processed by the distribution system according to its position in the building. A control room is located in the roof structure of the SF.

End-effectors As in the SMART all trolley hoist manipulators in the Hybrid-SMART are equipped with multipurpose hooks, which are lowered/lifted by two ropes from the trolley manipulator's main mechanical part. The simplicity of the end-effectors is accomplished by ROD.

System Variations The SMART-Hybrid is composed of elements of the SMART stating a fully modular kit. The automated part of the Hybrid-SMART (covered by the SF) can be used to construct rectangular building core areas with an orthogonal alignment of components. The sections of the building arranged around the core area are built by the not automated part of the Hybrid-ABCS and can be of any shape; orthogonal alignment of components would be advantageous but is not a prerequisite for the use of the system. The SF can be equipped either with the Trolley Hoist OMs of the SMART or with with simplified gantry-type OMs.

Robot-Oriented Design On a component level, the use and operation of the OMs is simplified by compliant and self-adjusting column joints (rack and pinion system). On a building level, the structure of the overhead crane system has to be synchronized with the shape and dimensions of the building's core area. The core area, which is covered by the automated part (SF) of the Hybrid-SMART, has to cover on the one hand in terms of logistics and component supply through the automated part of manufacturing a strategically advantageous area of the building and on the other hand a rather complex section of the building, which justifies the use of the OMs.

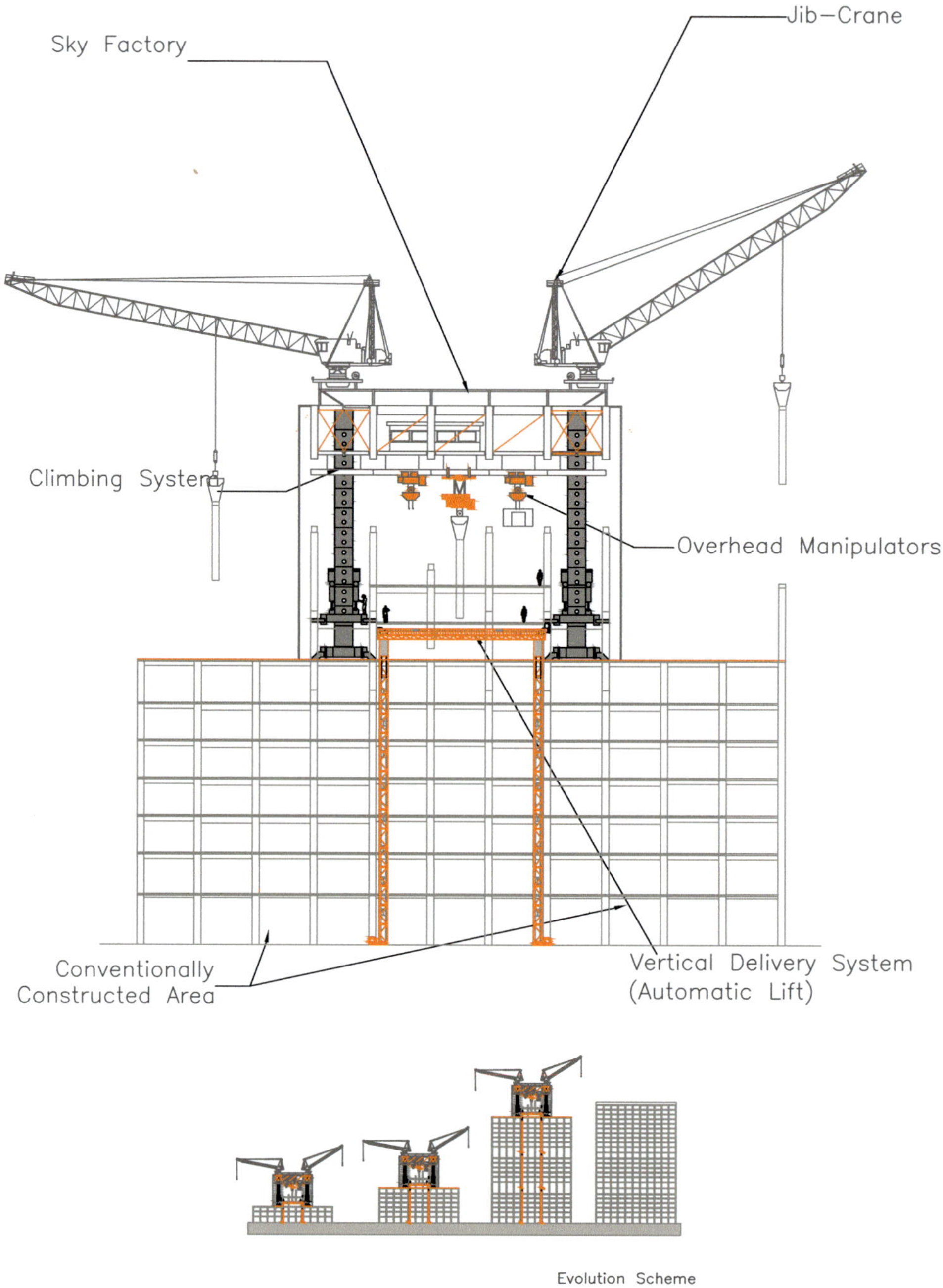

Hybrid Shimizu Manufacturing System by Advanced Robot Technology (Hybrid-SMART), Company: Shimizu Corporation, Japan

Figure 2.355. Sky factory, side view.

Figure 2.356. Sky factory, front view.

Figure 2.357. Ground factory.

Figure 2.358. Sky factory, interior view.

Figure 2.359. Installation of column components.

Figure 2.360. Installation of prefabricated sanitary modules.

Figure 2.361. The sky factory completely encloses the working space.

Figure 2.362. Installation of prefabricated damping elements.

Figure 2.363. Sky factory, roof view.

2.10 Decentralized Sky Factory (moving upwards) in Combination with Conventional Construction

Systems in this category are characterized by the application of decentralized and loosely coupled subsystems of SFs. So far all systems within this category have been the result of the phenomenon of technology diffusion, meaning that highly automated SFs were split into their subsystems and that the subsystems were used independently from each other and not combined to fully integrate automated supply chains such as in the original versions. Both Obayashi and Shimizu use those subsystems (e.g., automated component lift, automated welding systems, etc.) in conventional construction sites along with conventional work processes. The decentralized and loosely coupled use of the automated subsystems accounts for flexibility and low machine costs on the one hand, but on the other hand this strategy is unable to achieve high productivity in terms of time reduction or a reduction in human labour.

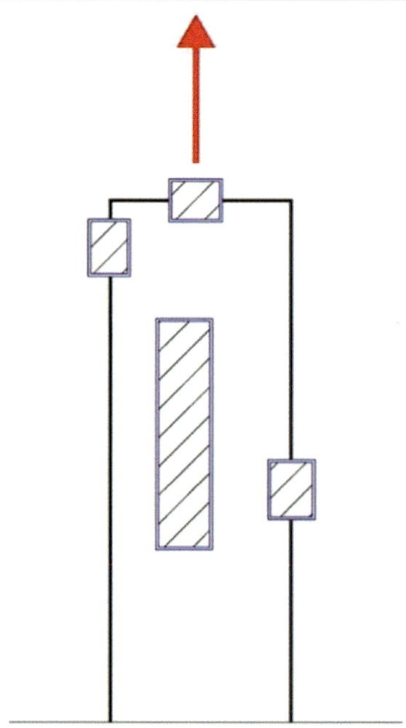

	System name	Company
23	Loose Deployment of ABCS-Subsystems (e.g., Tokyo Skytree)	Obayashi
24	Loose Deployment of SMART-Subsystems (e.g., Mode Gakuen Cocoon Tower)	Shimizu

Two recent and well known examples of the application of SF subsystems in a decentralized manner are the erection of the Tokyo Skytree (here Obayashi applied

ABCS and Big-Canopy subsystems) and the erection of the Mode Gakuen Cocoon Tower in Tokyo (high-rise building in Tokyo; here Shimizu applied subsystems of the SMART). Apart from these two projects, both companies apply subsystems of their automated site systems frequently in conventional construction projects. The application of subsystems relates the technological development in the field back to the idea of single-task robots (described in **Volume 3**). Individual automated technologies are coupled to cooperating systems that are compatible with conventional construction processes, but that are not as rigidly interconnected and automated as full versions of automated sites. As the application of subsystems provides advantages in terms of flexibility and cost it is highly likely that in the future individual subsystems or sets of subsystems will be developed for loosely coupled and decentralized use.

2.10.1 An Example of a Decentralized Subsystem Application of Automated Building Construction System (ABCS subsystems)

Company: Obayashi Corporation, Japan
Category: Decentralized SF (moving upwards) in combination with conventional construction

Analysis of Technology and Composition

Evolution Scheme Today, Obayashi uses subsystems developed for the use in ABCS more frequently as loosely coupled elements and not as centrally integrated elements within on-site construction. The use of subsystems of ABCS was not foreseen when ABCS was developed in the 1990s. The use in, and fusion of, highly complex ABCS

Figure 2.364. TOKYO SKYTREE™ was constructed using ABCS subsystems. (Image: TOKYO SKYTREE™/Obayashi)

technology with conventional construction methods and technologies can be seen as a form of technology diffusion. Obayashi recently applied ABCS subsystems in a loose combination to support the construction of the Tokyo Skytree.

Elevation Using and adjusting ABCS subsystems was, in the case of the Tokyo Skytree, highly efficient, as the 634-metre tall structure is an extreme project requiring high precision and technical support (Figure 2.364). Obayashi used prefabricated, large-scale steel elements to construct the Tokyo Skytree. As the Tokyo Skytree is extremely high, it has to behave exactly as predicted in the event of an earthquake, and a high-level of precision is necessary in component alignment. At extreme heights, welding support is also necessary, as well as an efficient parts and components delivery system. Obayashi therefore decided to use adjusted versions of various ABCS subsystems (subsystem for alignment and accuracy management, subsystem for vertical logistics automation, subsystem for welding automation).

Ground Plan In the event of the loose combination of ABCS subsystems, the idea of a factory which covers the building completely (ABCS) or partly (Hybrid-ABCS) is abandoned and subsystems are applied in a more loose combination. With the abandoning of the factory structure, OMs and CSs are also abandoned. All other subsystems become optional elements that can be used individually or in a combination, and are integrated by the management of an RTMMS. However, in many cases the vertical logistics subsystem (automatic lift) plays a central role.

Subsystems The optional subsystems that can be used in a loose combination in a decentralized factory are listed below. As the combination is loose, other machines and systems with network capability can simply be added. The list thus only shows the basic subsystems but not the complete possibility of the system.

1. *Vertical delivery system (optional, but basic)*. Being challenged by the highly compact and crowded central areas of Tokyo, which do not allow major disturbances by construction, Obayashi, like other major Japanese contractors, has developed JIT, JIS type supply and construction methods. A set of such methods was also applied in the construction of the Tokyo Skytree. Inside the steel frame core, an ABCS subsystem for automated logistics was installed to make the site logistics more efficient. This installation pays off especially in the building of a 634-metre high tower.
2. *Facade element delivery and installation system (optional)*. Under the factory roof, along the facade (outside of the floor area but along the building facade/edges), a rail-guided crane manipulator system can be installed for lifting and positioning of facade elements.
3. *Welding robot system*: *welding*. At a height of 634 metres (such as with the Tokyo Skytree), even the smallest aberration in the structure can weaken it and hinder the alignment of another large-scale prefabricated segment in subsequent construction phases. When the high-precision alignment and joining of steel

segments has to be guaranteed, Obayashi's automated welding systems are a major advantage. The welding system precisely and simultaneously welds elements around the joint from opposite sides. Welding from opposite sides, or in a specially assigned sequence, can guarantee that the joints and thus the segments do not distort during the welding process. The ABCS welding subsystem is controlled by welding software and allows the simultaneous welding by multiple robots on one joint.

4. *Mobile (horizontally/floorwise operating) logistics robots*. On floors within the SF that are already covered with ceiling elements.
5. *Alignment and accuracy measurement system*. The construction of the 634-metre high Tokyo Skytree structure required a high degree of precision to achieve the desired performance. To achieve this, the ABCS subsystem for automated and high precision column alignment of steel structures was used. This system had been improved for the construction of the Skytree.
6. *Real-time monitoring and management system*. Data from sensor systems, as well as from the servomotors/encoders of the V/HDS, are used for management and real-time control of the construction process.
7. *Further subsystems*. As the combination is loose, other machines and systems with network capability can simply be added. In the case of the Tokyo Skytree an automatically rinsing/climbing and adjustable concrete slip form was added as subsystem to be able to automate the construction of the inner concrete core of the Skytree.

End-effectors As the OM system is not applied, the system of dedicated, component adjusted end-effectors that is closely related to these systems is also not applied. However, in the case of the Tokyo Skytree, for example, Obayashi developed a specialized (nonrobotic) end-effector system that was attached to the conventional tower cranes that rose with the Skytree on two sides (Figures 2.365 – 2.369). The end-effector system allowed the efficient picking up and positioning of the prefabricated, large-scale steel elements.

System Variations The loose and decentralized combination of subsystems allows the ABCS application to be almost independent from column grid structures, and also from vertical or horizontal orthogonality. In the case of the Tokyo Skytree, this flexibility was necessary, as the building's form and composition frequently changes from bottom to top.

Robot-Oriented Design On a component level, the size, dimension, and weight of components has to be designed to fit the requirements for use by the subsystems (e.g., automatic lift). In addition, on a building level, the allocation of the vertical logistics subsystem needs to be decided. In the case of the Tokyo Skytree, the inner, concrete-based round core, determined the position of the automatic lift.

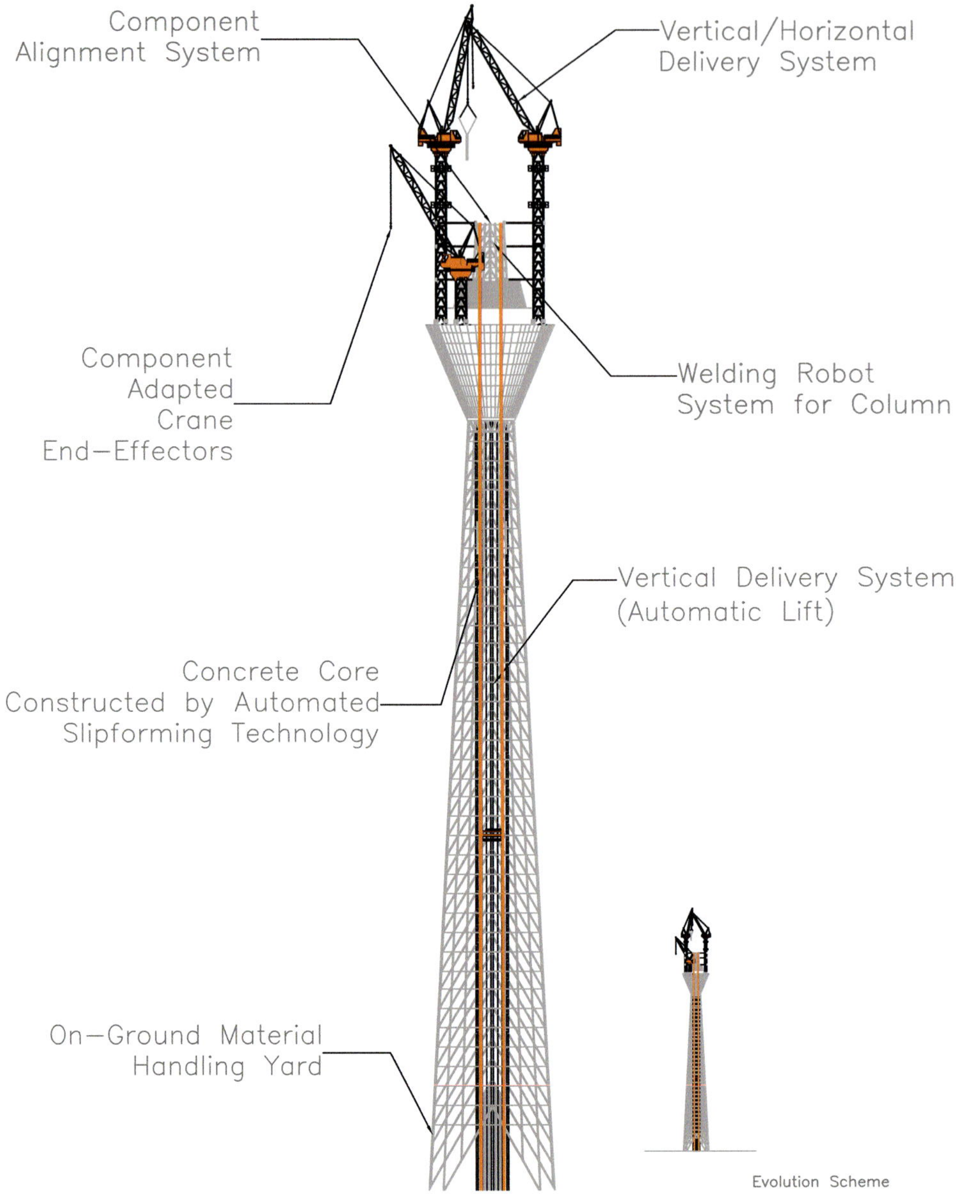

Decentralized Subsystem Application of ABCS (ABCS subsystems), Company: Obayashi Corporation, Japan

Figure 2.365. Self-climbing cranes. (Image: TOKYO SKYTREE™/ Obayashi)

Figure 2.366. Structure assembled by use of component alignment system. (Image: TOKYO SKYTREE™/ Obayashi)

Figure 2.367. Specialized end-effector for component pickup/positioning. (Image: TOKYO SKYTREE™/ Obayashi)

Figure 2.368. Stilts of the self-climbing cranes. (Image: TOKYO SKYTREE™/ Obayashi)

Figure 2.369. Specialized end-effector for component pickup/positioning. (Image: TOKYO SKYTREE™/ Obayashi)

2.10.2 An Example of a Decentralized Subsystem Application of Shimizu Manufacturing System by Advanced Robot Technology (SMART subsystems)

Company: Shimizu Corporation, Japan
Category: Decentralized SF (moving upwards) in combination with conventional construction

Analysis of Technology and Composition

Evolution Scheme Shimizu used concepts, techniques, and subsystems that were initially developed for the concentrated application within SMART in a loosely coupled and only partly automated form to construct a high-rise building in Tokyo (Figure 2.370). In a temporary GF, parts and components were transformed into higher level components (steel structure components, facade components) to minimize assembly work for the complex structure on top. The high-level steel structure components were assembled in the SF, while on the lower levels the facade components were attached by tower cranes equipped with component adapted, specialized balancing end-effectors.

Elevation The steel main structure was assembled on the uppermost floors. A factory structure providing side cover and working platforms to support fixation and welding operations climbed up the assembled structure. A vertical lift was installed within one of the three main shafts, which delivered the structure and facade components to the top to be picked up by the tower crane manipulators. The vertical lift picked the components up from a temporary GF installed on the ground level cafeteria part of the building. A chain-like manufacturing sequence was established on-site for the on-site transformation of facade parts into high-level facade components.

Ground Plan Around the building's central core, the steel structure of the building created three vertical shafts. Extendable tower crane manipulators were positioned in two of the shafts; in the third shaft, a vertical lift was installed that delivered components from the GF to the operational floors at the top of the building.

Subsystems The decentralized site-manufacturing setting (system) consists of many (integrated) subsystems:

Figure 2.370. Climbing cover and scaffolding structure.

1. *Climbing cover and scaffolding.* A climbing cover and scaffolding structure spanning two floors climbed along the structural components located on the top floor (Figures 2.371–2.372). It provided a weatherproof and safe environment, as well as outside working platforms for final fixation and welding.
2. *Vertical delivery system.* In the third shaft, a vertical lift was installed, delivering components from the GF to the operational floors on top of the building. Multiple components (e.g., up to six facade components) could be delivered at once.
3. *Vertical/horizontal installation system.* In two of the shafts, extendable tower crane manipulators were positioned, which were capable of being equipped with different end-effectors.
4. *Welding robot system (optional).* Shimizu also had the option of using its column attachable welding robots within the weatherproof climbing cover and scaffolding.
5. *Ground factory.* For the on-site transformation of facade parts into high-level facade components, a chain-like manufacturing sequence was established on-site (delivery of parts, joining of delivered frames to larger elements, installation of glass and random strips, sealing). Components in the temporary GF were assembled JIT and JIS, according to the installation sequence in an SF. Software was used to simulate and optimize the assembly sequence.
6. *Real-time monitoring and management system.* Software was used to simulate the construction process, and to optimize the assembly sequence of the elements of the complex not-rectangular facade in particular.

End-effectors As was demonstrated with the high-rise building in Tokyo, the vertical/horizontal installation system can be equipped with various end-effectors suiting the pickup, positioning, and adjustment of various component types:

1. End-effector for pickup, positioning and adjusting of structural steel components
2. End-effector with a balancing device for orientating and positioning facade components

System Variations The construction of the high-rise building in Tokyo has shown how the loose coupling of subsystems initially developed for the SMART can allow a complex shaped building, in which nearly every floor, structural element, and facade element is different, to be erected in a structured and systemized on-site environment.

Robot-Oriented Design The structural steel structure surrounding the inner core and shafts of the high-rise building was modularized into a kit of columns, beams, and corner components with joints that simplified temporary fixing after the positioning. Even though the facade seems to have an irregular structure, the facade was modularized into rectangular elements that could be assembled in the GF, and that simplified the installation process.

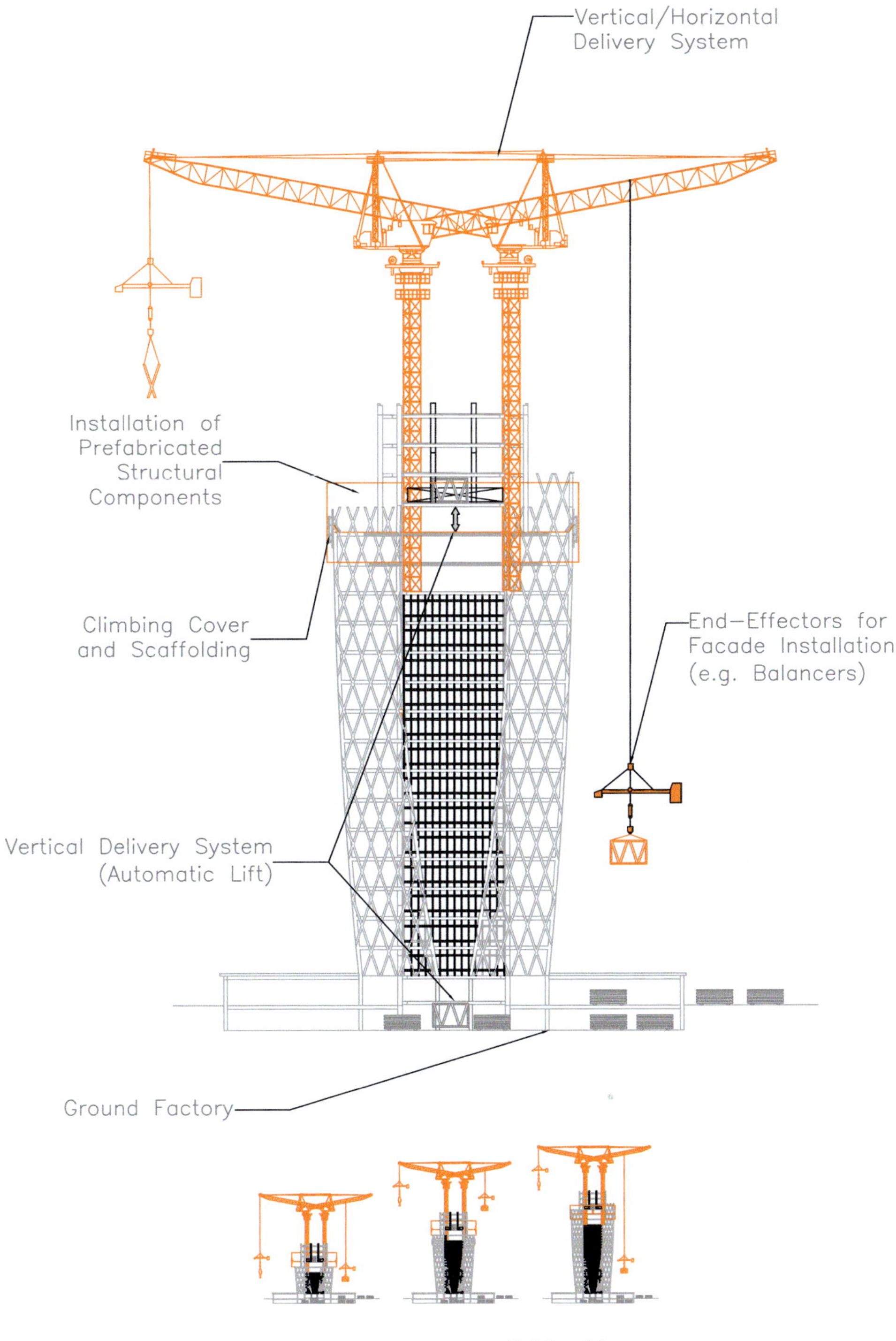

Decentralized Subsystem Application of Shimizu Manufacturing System by Advanced Robot Technology (SMART subsystems), Company: Shimizu

Figure 2.371. The climbing cover and scaffolding structure partly structures the assembly environment on-site, early construction stage.

Figure 2.372. Climbing cover and scaffolding, side view.

Figure 2.373. Climbing cover and scaffolding, detailed view.

3 Analysis and Categorization: Deconstruction

Since 2008, when Kajima started the deconstruction era, the six major Japanese contractors had developed mechanized and partly automated deconstruction systems. Deconstruction systems follow the same approach as automated construction sites and install an on-site factory (fixed type or moving type), which then provides the basis for controlled, structured, and systematized on-site work processes. The cover of the on-site working area provides – apart from the advantage of creating a weatherproof and disassembly supportive environment (with the possibility of installing overhead cranes, cameras, and other measuring technology) – especially in building deconstruction, the advantage of being able to reduce the noise and dust disturbance to the surrounding environment (e.g., in the case of the TECOREP system, noise disturbance for the surrounding environment was reduced to less than 20 decibels). Deconstruction sites are work environments where processes are, in a factory-like manner, phased to a strict time schedule, and the flow of material is highly organized. Deconstruction sites thus follow the general idea of a structured environment (SE).

Deconstruction in Japan is a thriving market. As a result of to the growing population density, a huge number of high-rise buildings have been constructed since the early 1990s. Many of those high-rise buildings are no longer considered to be 100% fail-safe, owing to advanced knowledge of structural design and to a change in thinking and requirements after the March 2011 earthquake (magnitude: 9.0) and the predicted increase of earthquakes in the future. Developers and owners of those high-rise buildings thus have to consider either the (costly) integration of damping systems or the complete deconstruction and rebuild. The second method also has the potential benefit that during a rebuild, buildings can also be adapted to new functional, technological, and ecological requirements. This requires, however, that the deconstruction and rebuild can be done faster than with conventional methods. As the deconstruction and rebuild of high-rise buildings is a highly profitable business in Japan, and a huge amount of combined deconstruction/rebuild orders are expected over the next decade, major Japanese contractors are competing over deconstruction ability and speed.

A further advantage of systematic or automated deconstruction – especially in the Japanese economic environment – is that, apart from the high speed of deconstruction, the achievement of (compared to conventional demolishing) an extremely high rate of recyclability or reusability of the deconstructed materials, parts and components (in the case of a DARUMA application in Tokyo, the recycling rate exceeded 90%). The deposit of construction material in Japan is highly expensive, as because of the country's geographical features (limited island area, many hills and mountains, large wooded areas, limited possibility of creating habitable space), the population density is high and construction waste represents a threat to the possibility of using this space efficiently. In addition, since about 2005, the Japanese government has been promoting making the construction sector more sustainable (houses ought to be able to last 250 years, resource and energy consumption need to be reduced). Since the 2011 Tohoku earthquake, and related Fukushima nuclear power plant incident – which led to a permanent shutdown of nearly all 50 nuclear reactors in Japan, the saving of resources, as well as the change towards ecological thinking, has become a key issue in Japanese society, industry, and science.

Most deconstruction sites first disassemble the larger (high- or low-level) components, then disassemble those components (e.g., in a ground factory [GF]-like setting) on-site into lower level components, mono-material parts, or raw materials that can then be delivered directly from the site to the recycling plant or enterprise. Furthermore, the research departments of some companies (e.g., Kajima) are currently seeking ways to convert deconstructed material directly back into usable building material on-site. Similar to automated construction, central to the on-site demanufacturing process are, apart from the implementation of a controlled environment and controlled material flow, the use of overhead crane systems, vertical logistics lifts, and sensor and measurement technology for computer-supported monitoring of the construction process (progress monitoring, decibel monitoring, dust and CO_2 ratio monitoring).

Deconstruction sites are operated – just like automated construction sites – on a 4–7-day cycle (time planned for deconstruction of a single floor). The automation ratio of deconstruction sites is (to date) lower compared to automated construction sites, and many subsystems are remote controlled and human supervised. This is due, on the one hand, to the higher flexibility required in deconstruction and, on the other hand, to the lower implementation costs necessary. However, as more and more companies start using deconstruction systems, the pressure to improve the existing systems increases, and companies are likely to improve the methods and technology through repeated use. When contractors are asked in the future to deconstruct buildings that were actually constructed using automated technology, the possibility of using previously generated systems and data sets for fully automated deconstruction could be explored.

Most companies are also aiming at, apart from the aforementioned advantages, higher productivity and the reduction of human labour. Systems in the following section that are categorized under category 1 or 3 can potentially reduce the amount of human labour required. Systems in category 2, on the other hand, aim less at significant cost and time savings, but rather focus on addressing the advantages of increased safety, a higher recycling rate, and less disturbance to the environment. Taisei, which has so far deconstructed buildings of up to 105 metres high, has stated that highly time and cost-efficient deconstruction could be accomplished by TECOREP when the building to be deconstructed is higher than 150 metres. Basically, for deconstruction, the same analysis framework as for automated construction sites could be applied. Deconstruction sites also use climbing subsystems, factory subsystems, logistics/crane subsystems, and subsystems to control the deconstruction in real-time. New or different to automated construction are the deconstruction subsystems for the real-time supervision of noise and dust/dirt levels. Furthermore, deconstruction required some new types of end-effectors (e.g., material sorting for recycling, water-cutting, laser-cutting).

3.1 Closed Sky Factory Supported by Building (moving downwards)

To date, two deconstruction systems have been developed in this category. Systems in this category are characterized by the installation of a closed sky factory (CSF) on top of the building. Within the CSF, the building is deconstructed from top to bottom. Similar to automated construction, the lifting mechanism (in the case of deconstruction this is better referred to as a lowering mechanism) that allows the lowering of the factory after each floor cycle is key. Two variations of lowering mechanisms in this category can be identified so far:

- The Hat Down factory is lowered by a mechanism that encompasses the building from outside and penetrates the facade at certain clearly defined points, to get a grip on structural elements.
- The TECOREP factory is lowered by a mechanism that vertically penetrates the inner floors at certain clearly defined points, and that also uses the floor structure to get a grip.

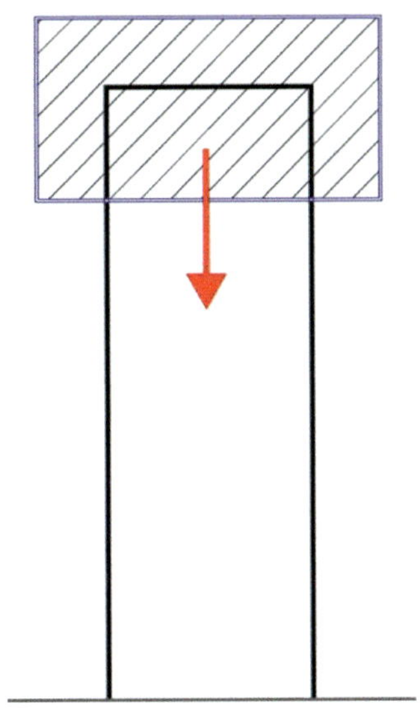

	System name	Company
25	Hat Down	Takenaka
26	Taisei Ecological Reproduction System (TECOREP)	Taisei

The Hat Down factory is a completely new structure assembled on-site. The TECOREP factory utilizes parts of the building that are to be deconstructed and uses one of the top floors as the basic structure and roof of the factory, to which it assembles the overhead cranes and the side covers. All systems in this category disassemble larger elements in the sky factory (SF), with additional disassembly on the ground floor.

3.1.1 HAT Down

Company: Takenaka Corporation, Japan
Category: Deconstruction – closed sky factory supported by building (moving downwards)

Evolution Scheme Takenaka Corporation developed the Hat Down method, which allows the disassembly of high-rise buildings in the centres of cities and other complex locations by installing a movable disassembly on-site factory (the CSF; Figure 3.1) so that the top of the building is enclosed. It then climbs down as the building is disassembled from top to bottom. The CSF, in which the equipment necessary for disassembly is integrated, including overhead manipulators (OMs) and other subsystems, completely encloses the building during lowering. All the demolished materials

Figure 3.1. Deconstruction by the sky factory.

are lowered through shafts inside the building, allowing a disassembly process that is safer and freer from disturbance to the surrounding than conventional methods.

Elevation The outside of the CSF is completely covered with soundproof panels and its ceiling is designed so that it could be opened and closed, depending on the weather conditions. Twenty-two jacks are installed to raise and lower the CSF, and after the interior of the floor where the CSF is installed is disassembled, the entire CSF is lowered down to the next floor in about 1 hour to prepare it to disassemble the next floor. In parallel, at the bottom of the building, all the lowered elements are processed to separate the different materials for recycling or re-use on other construction sites.

Ground Plan The floor to be disassembled is divided into several sections, each equipped with an OM and a lowering opening. Workers work simultaneously in all sections, disassembling and cutting the columns, walls, floors, and so forth into about 200 high- to medium-level components per floor, and then lowering these to the ground floor through the lowering shafts. The bottom four floors are cut into pieces with heavy machinery in the SE of the CSF orderly. The Hat Down system works better when applied to buildings with a big difference between their length and width as this allows for an optimal utilization of the OMs and the parallel processing in various factory parts. To apply the method to a building with a more complex geometry, modifications would be necessary. In addition, it is useful to find existing openings to be used as lowering openings for the system, such as elevator shafts.

Subsystems The HAT Down system consists of a set of integrated subsystems:

1. *Sky factory roof structure*. A lightweight steel frame structure spans the building and provides a platform for the installation of the gantry type OMs (Figures 3.1–3.3, 3.8). The cover of the CSF consists of soundproof panels to minimize a disturbance of the surrounding area by the disassembly work. The CSF has an adaptable ceiling that can be opened and closed depending on the weather condition. Solar panels are located on the CSF's outside to produce electricity.
2. *Climbing system*. The system is supported by two end-effectors (grippers; Figure 3.6) that hold themselves to each column of the building using compression (Figure 3.7). The structure is lowered in a computer-controlled way by a screw mechanism (Figure 3.5) via the installation of a third gripper one level below. The gripper on top is then removed and a new level is ready for disassembly. The end-effectors of the CS also function as template for cutting and disassembly (Figures 3.9–3.11).
3. *Horizontal delivery system*. Each of the three zones has a gantry type OM (Figure 3.4) to transport the demolished components down to the ground level. OMs can move in the three directions *X–Y–Z* along the main beams of the system, allowing them to cover the whole workspace.
4. *Lowering shafts*. Lowering shafts (Figures 3.12 and 3.13) allow the OMs to move all the materials down to the material handling, sorting, and processing yard (MHSPY). It is possible that already existing elevator shafts can be utilized as lowering shafts.

5. *Material handling, sorting, and processing yard.* An MHSPY is installed in the ground floor of the building for disassembling of components and sorting of materials (Figures 3.14 and 3.15). Furthermore, materials are prepared in the MHSPY to be picked up by factory external logistics (FEL). The pickup of components by FEL can thus be organized just in time (JIT) and just in sequence (JIS).
6. *Real-time monitoring and management system.* A series of sensors are deployed all over the site to allow a strict internal control (Figure 3.16). Variables such as electric current and voltage in the equipment are monitored and controlled to raise performance, quality, and safety. Aspects such as temperature, humidity, noise levels, and CO_2 emissions are also controlled to maintain the best working conditions.
7. *Templates for cutting.* When demolishing the columns of the building, the gripper of the climbing mechanism is, for example, used as a template for making a clean cut at the base of the columns (Figures 3.9–3.11). In this way a standardized size of disassembled column elements is ensured.

End-effectors

1. *Climbing end-effectors.* They are part of the climbing system and transmit all the weight of the CSF to the columns of the building. The end-effectors consist of two identical elements placed on opposite sides of a column. Using bolts, the two elements are pulled towards each other to hold them in place on the column.
2. *HDS end-effectors.* The HDS is equipped with lockable hook end-effectors, whose position is determined by the HDS. The demolished elements are attached to the hooks by the workers using a set of ropes and steel cables that are customized and preconfigured for the elements to be disassembled for a certain building.

System Variations

- *Longitudinal expansion.* The system is applied mainly to rectangular buildings with one considerably longer dimension. This is because the HDS cannot span more than 25–30 metres. However, the system can be easily expanded in the other direction (lengthwise) and would in modified form also be applicable to horizontally oriented buildings.
- *Multidirectional expansion.* This system modification could be implemented if the internal columns of the building are inside the 25–30-metre HDS span and supported by the HDS's workspace.

Robot-Oriented Design So far ROD was not considered in the design of the building. However, in the context of the HAT Down system the following design features would make the application of the system more efficient:

- On a parts and component level, the number of different types of materials has to be minimized to simplify the process of processing materials on-site and reduce transport to separate reprocessing plants.
- Connections that can be mechanically separated should be used (instead of welding and gluing).

- A minimum number of different types of connectors should also be used to require fewer types of tools and equipment in the CSF.
- A structural grid should also be used that is fully in tune with the structural grid of the CSF.
- Integrated information modelling that assigns to the individual parts and components properties that are relevant for disassembly could build the basis for Agent-Based Modelling (ABM) and an optimized scheduling and control of the (automated) disassembly process (for further information, see Lee et al., 2015).

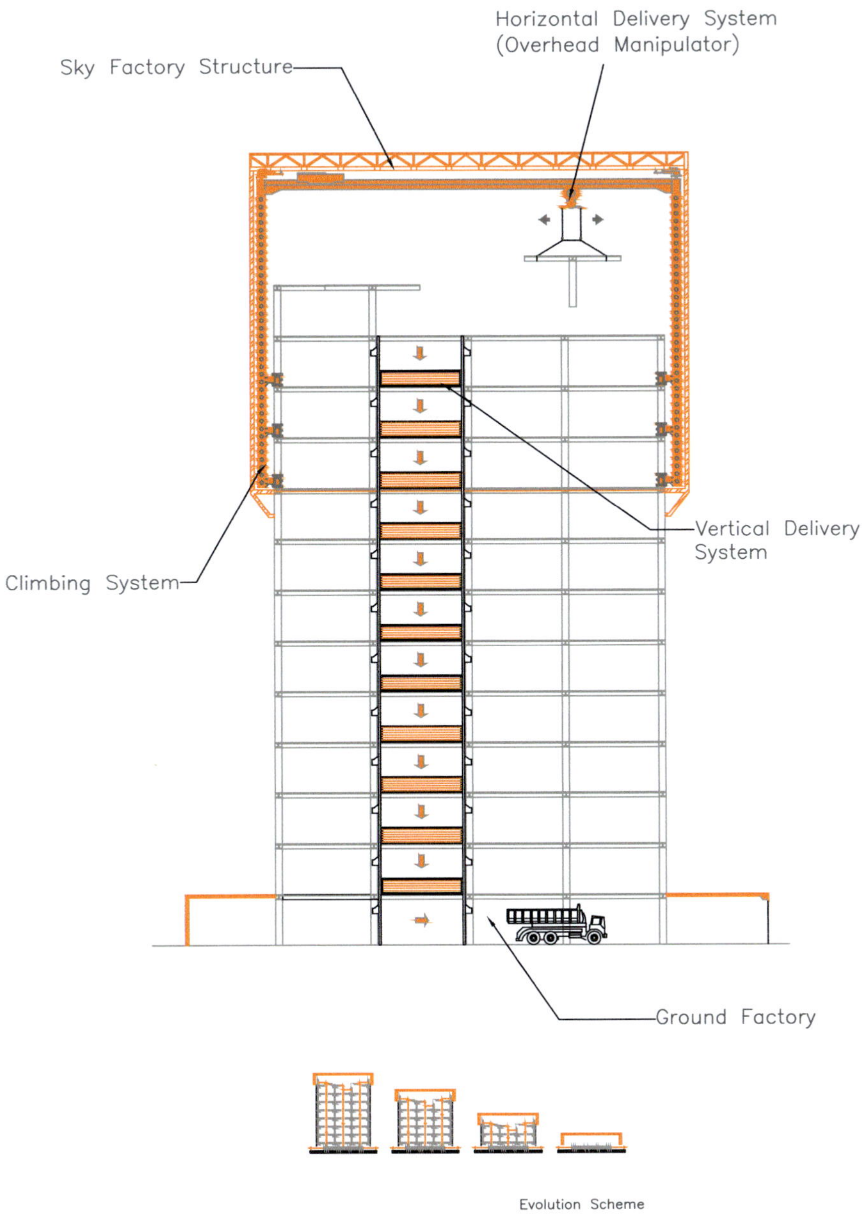

HAT Down, Company: Takenaka Corporation, Japan.

Figure 3.2. The sky factory completely encloses the working space in which deconstruction activity takes place.

Figure 3.3. Sky factory, interior view.

Figure 3.4. Overhead manipulator.

Figure 3.5. Climbing mechanisms.

Figure 3.6. End-effector/gripper of the climbing system.

Figure 3.7 Stilts of the climbing system are fixed to the building by end-effectors.

Figure 3.8. Structural components are prepared for disassembly by laser cutting.

Figure 3.9. The end-effectors of the climbing system in the uppermost floor serve as cutting templates.

Figure 3.10. Laser cutting.

Figure 3.11. Removal of structural component.

Figure 3.12. Disassembled components are lowered to the ground factory through lowering openings (vertical shafts).

Figure 3.13. Overhead manipulator lowering a disassembled component.

Figure 3.14. Entry gate ground factory.

Figure 3.15. Material sorting in ground factory.

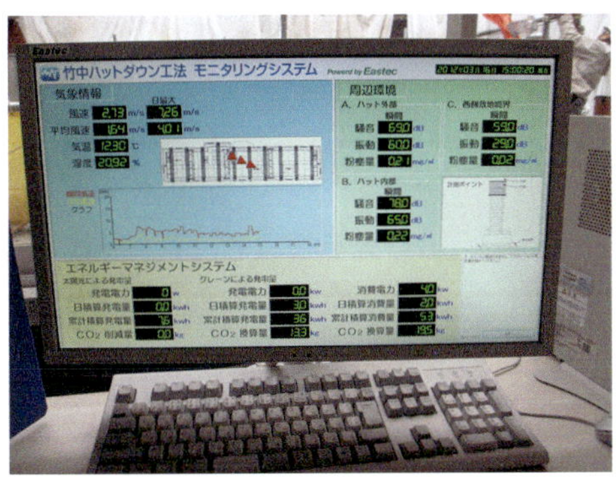

Figure 3.16. Real-time monitoring of construction process in on-site control room.

3.1.2 Taisei Ecological Reproduction System (TECOREP)

Company: Taisei Corporation, Japan
Category: Deconstruction – CSF supported by building (moving downwards)

Evolution Scheme TECOREP uses the roof of the building as the roof of the CSF. The lifting/lowering mechanism is installed and the columns of the uppermost floor

Figure 3.17. Sky factory used for systematic deconstruction.

are cut. The roof of the building is connected via the lifting/lowering mechanism and side shelters covering the workspace and providing working platforms for facade work, and a rail-guided OM system is attached to it. Further vertical shafts are created, through which disassembled parts and components can be lowered to the ground floor, where material is handled and sorted.

Elevation The climbing mechanism "breaks" through the floor slabs floor by floor, lowering the SF (Figure 3.17) after the uppermost floor has been disassembled. Work is done on several floors in parallel. On the floors under the uppermost operational floor, the breakthrough of the climbing mechanism is prepared, and the interior finishing is disassembled. On the uppermost operational floor, the main structural components of the building are cut off and then picked up and lowered by the OMs.

Ground Plan Compared to open sky factories (OSFs), TECOREP provides not only the advantage of creating a relatively independent environment to reduce the disturbance to the surrounding environment, but also the possibility of installing overhead cranes, cameras, and other measuring technologies. The CSF method is the best solution for the outside environment; noise disturbance can be reduced to lower than 20 decibels. All operations are monitored and supervised form a control room (Figure 3.24).

Subsystems The TECOREP system, which systematically disassembles buildings on-site, consists of many integrated subsystems:

1. *Sky factory roof structure*. The basic element of the SF is the roof of the building to be deconstructed, which covers working platforms, the climbing mechanism, and is the element to which OMs are connected (Figures 3.18, 3.19, and 3.25–3.28). The SF completely covers the building and provides a weather-immune SE, while lowering the sound and dust disturbance to the environment (Figure 3.23).
2. *Climbing system*. The CS equals that of Takenaka's Hat Down in its composition. The CS consists of two main parts: core and frame. On the core there are three column grippers. The grippers grasp the column over three different storeys. The frame also has three grippers, which are placed on the floor slab to secure the system's position. However, in contrast to Takenaka's Hat Down, the climbing mechanism does not operate along the facade, but from within the floor area itself, therefore having to break through floor by floor. The advantage is that the facade does not need to be prepared for the attachment of the climbing mechanism.
3. *Vertical/horizontal delivery system*. Automatic rail-guided overhead gantry type OMs are installed to the SF's roof (Figures 3.20 and 3.21). The manipulators are used to hold components during disassembly (such as cutting them off), to pick them up and transport them horizontally over a vertical delivery shaft, and to lower them through the shaft down to the ground.

4. *Material handling, sorting, and processing yard (part of ground factory).* In an MHSPY, the disassembled components lifted down from the SF are disassembled further; material is sorted and prepared for transportation. Through preparation and sorting, the amount of material to be disposed of is reduced to a minimum.

End-effectors The OMs are equipped with multipurpose conventional crane end-effectors (Figure 3.22). Furthermore, to guarantee a precise disassembly, new methods for on-site water and laser cutting are applied and tested.

System Variations The installation of the SF and the CS do not require synchronization with the building's grid of columns. The subsystems can be installed to almost any building that allows that the roof structure to be used as a platform for mounting the described subsystems.

Robot-Oriented Design To date, no building code or guideline exists that, on being applied, would simplify the use of TECOREP.

Project 1

Figure 3.18. Installation of a sky factory. (Image: Taisei Corporation)

Figure 3.19. Deconstruction sequence. (Image: Taisei Corporation)

Figure 3.20. Interior view of sky factory. (Image: Taisei Corporation)

Figure 3.21. Overhead manipulators. (Image: Taisei Corporation)

Figure 3.22. Concrete slab cutting and disassembly in the sky factory. (Image: Taisei Corporation)

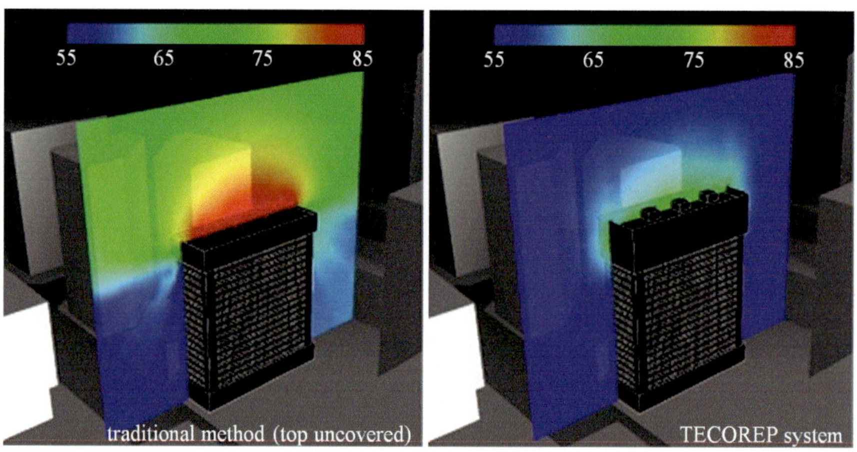

Figure 3.23. Noise level reduction. (Image: Taisei Corporation)

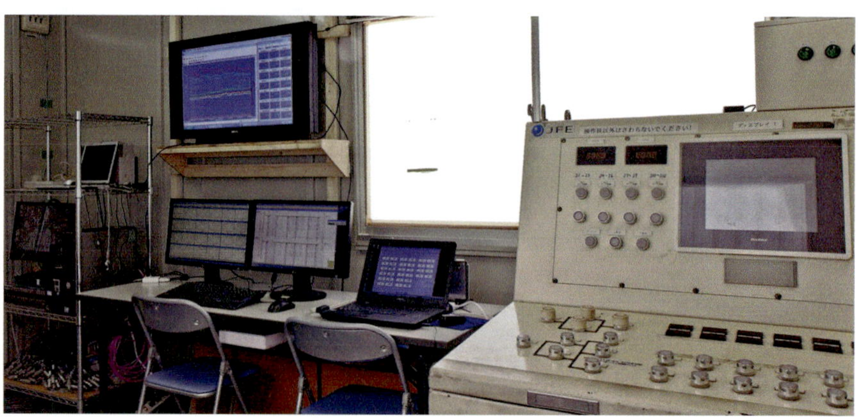

Figure 3.24. Control room. (Image: Taisei Corporation)

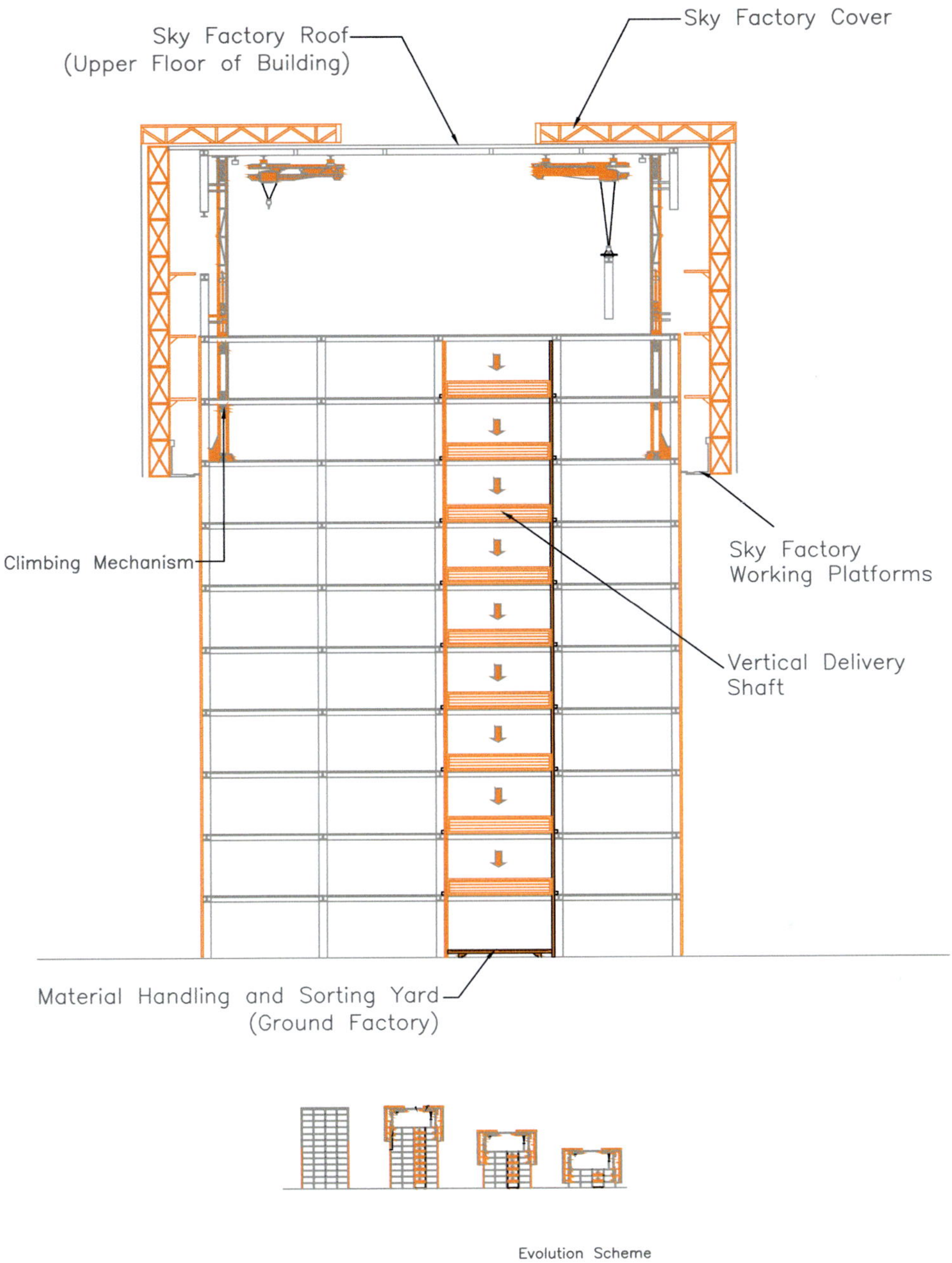

Taisei Ecological Reproduction System (TECOREP), Company: Taisei Corporation, Japan.

Project 2

Figure 3.25. Building after installation of sky factory.

Figure 3.26. The sky factory completely encloses the workspace.

Figure 3.27. The sky factory's cover smoothly follows the not-rectangular shape of the building.

Figure 3.28. Cover structure of the sky factory.

3.2 Open Sky Factory Supported by Building (moving downwards)

To date, three deconstruction systems have been developed in this category. Systems in this category are characterized by the fact that the building is deconstructed from top to bottom and that the roof of the SF is not completely closed and thus it is called an OSF. The factory only has side walls and encompasses the building like a closed band, moving downwards along the facade. The factory's main purpose is thus to limit the disturbance to the surroundings (aesthetically, and from noise and dust), and to provide assistance and platforms for dangerous disassembly work on the facade.

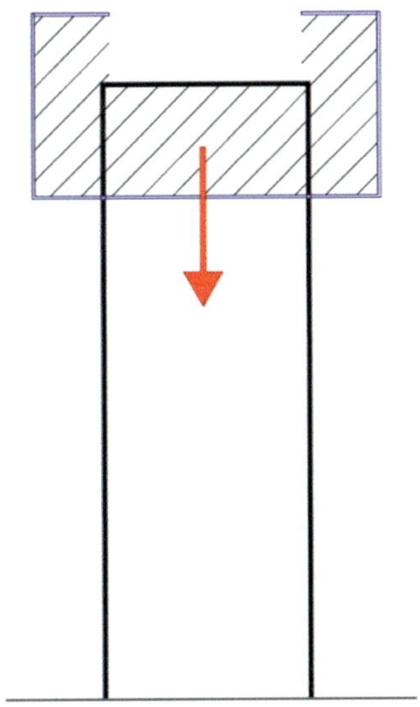

	System name	Company
27	Move Hat	Nishimatsu
28	Reverse Construction Method (RCM)	Shimizu
29	Quakeproof, Quiet, Quick and Block-by-Block (Cube Cut Method)	Obayashi

The pickup and lowering of components is done using standard cranes. On the one hand, the control and structuring effect of the work environment and logistics processes can be considered as being lower compared to the former category of systems with closed roofs. On the other hand, these systems are less capital intensive than the systems of the former category, and can thus be developed and implemented faster and at a lower cost. Systems in this category require almost the same time and cost as for conventional demolishing, but have the advantage of more safety, a higher recycling rate, and less disturbance to the environment.

3.2.1 Move Hat

Company: Nishimatsu Corporation, Japan
Category: Deconstruction – OSF supported by building (moving downwards)

Evolution Scheme The OSF (Figure 3.29) is set up on the ground floor and then lifted via the climbing mechanism to the top of the building (Figure 3.30), where it

Figure 3.29. Move Hat sky factory begins operation. (Image: RIBC, 2012)

partly covers the workspace and provides space for the operation of OMs, as well as safe working platforms and protection of the environment from noise and dust. The final disassembly and the breakdown of components into individual parts and materials are conducted by a GF.

Elevation The elevation shows the open roof structure of the SF and the working space of the OMs used to efficiently disassemble facade elements and outer structural elements. The SF encloses the workspace and provides the basis for structuring the working environment. It gives protection against the wind, sun, and rain.

Ground Plan Depending on the size of the building, a variable number of inchworm climbing mechanisms and vertical delivery systems (VDSs) can be installed. The open roof structure is modularized so that it can be adapted to a huge variety of orthogonally organized steel- and concrete-based buildings.

Subsystems The Move Hat system, which systematically disassembles buildings on-site, consists of many integrated subsystems:

1. *Open sky factory structure*. The SF of the Move Hat does not enclose the workspace completely but builds a covering ring around the top floors where the disassembly of structural elements is done. The SF does not span the whole building, but only the operational floors' external walls. The structure provides working platforms for efficient and safe disassembly work on the outside of the facade.
2. *Climbing system*. The CS is attached to the outside of the building and, via hydraulic actuation, climbs up/down the facade. As the system can climb up/down the facade, the set-up of the CS and the assembly of the factory side covers can be done conveniently on the ground.
3. *Partial horizontal delivery system*. The SF provides a roof for the installation of 3-DOF OMs only along the operational floors' sides (translation lengthwise,

translation sideways, Lower/lift). The overhead cranes can thus be used to support the disassembly of facade elements and of the outer structural beams and columns.

4. *Vertical delivery system*. Winch-based cranes are installed that lower the disassembled material via shafts created within the building's floor area.
5. *Lightweight equipment*. Lightweight, conventional construction machines can be used on the operational floor.
6. *Material handling, sorting, and processing yard (part of GF)*. The ground floors are also covered by a factory-like structure from the outside to prevent the disturbance of the environment by noise and dust. Components lowered from the SF are disassembled further on the ground; material is sorted and prepared for transport.

End-effectors The manipulators are equipped with multipurpose conventional crane end-effectors. However, to guarantee precise disassembly, new methods for on-site water and laser cutting are applied and tested.

System Variations The installation of the SF and the CS do not require synchronization with the building's grid of columns. The subsystems can be installed to almost any building that will allow the roof structure to be used as a platform for mounting the described subsystems. As the elements of the SF are orthogonal, the building should also be orthogonal to avoid major adjustments. The facade of the building to be deconstructed should be stable enough for the application of the CS.

Robot-Oriented Design To date, no building code or guideline exists, that, on being applied, would simplify the use of Move Hat.

Figure 3.30. OSF is set up on the ground and then lifted up the building by a climbing/lowering mechanism. (Image: RIBC, 2012)

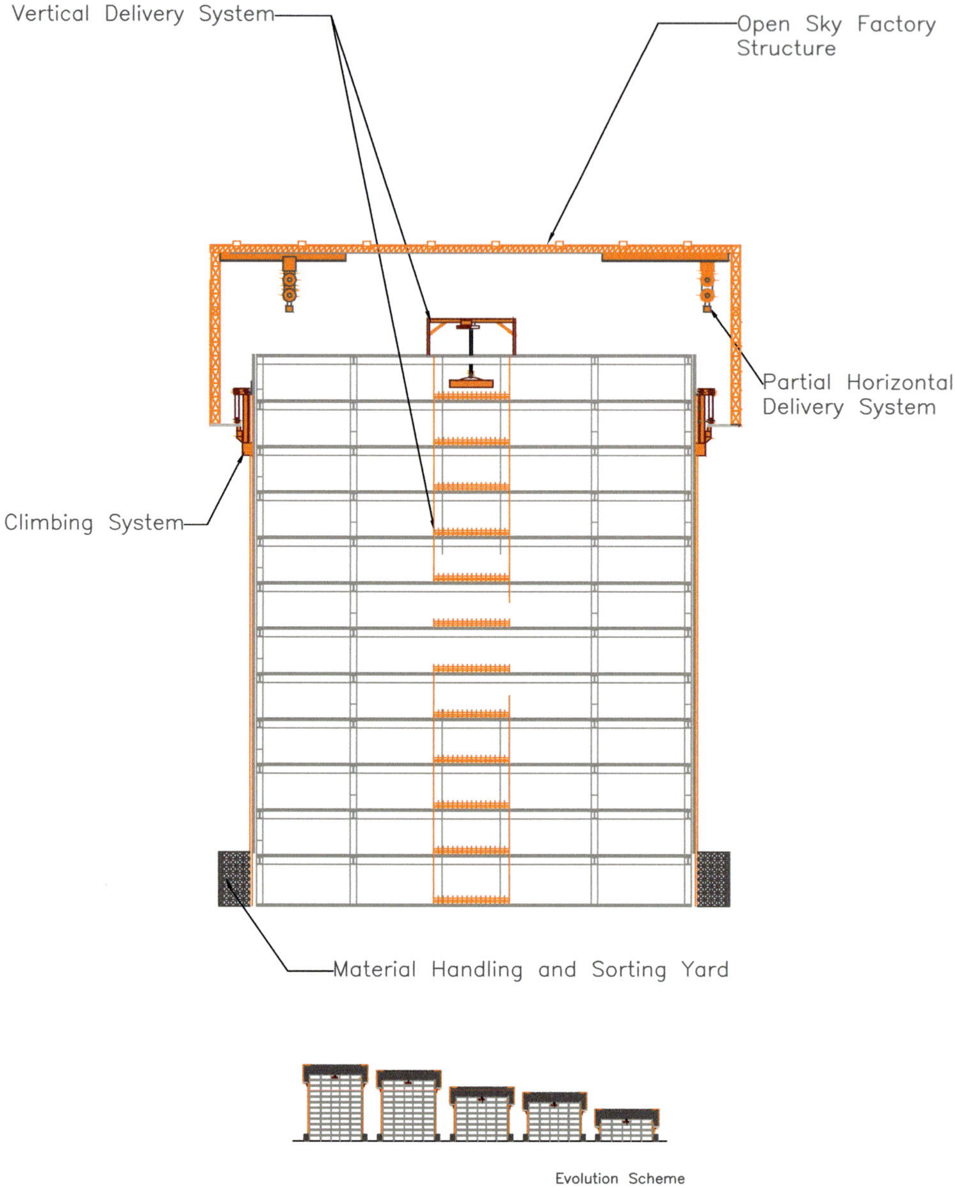

Move Hat, Company: Nishimatsu Corporation, Japan.

3.2.2 Reverse Construction Method (RCM)

Company: Shimizu Corporation, Japan
Category: Deconstruction – OSF supported by building (moving downwards)

Evolution Scheme The RCM SF (Figure 3.31) does not fully cover the workspace, providing only a movable protection ring around the top floors where disassembly is done. First, the interior plumbing elements are removed from the building (and

Figure 3.31. Open sky factory. (Image: RIBC, 2012)

vertical shafts created) and a tower crane capable of being lowered in parallel to the SF is set up. Second, the scaffold unit (OSF) is assembled on the ground and lifted up along the facade. Third, the building's shell is cut into blocks so that it is suitable for vertical logistics – deconstruction proceeds floor by floor. Finally, the OSF is disassembled on the ground.

Elevation In contrast to other systems, the RCM does not work with overhead cranes or lifts lowering the material within the floor or via shafts, but uses a tower crane as a central element, which can be lowered floor by floor in parallel with the SF. The

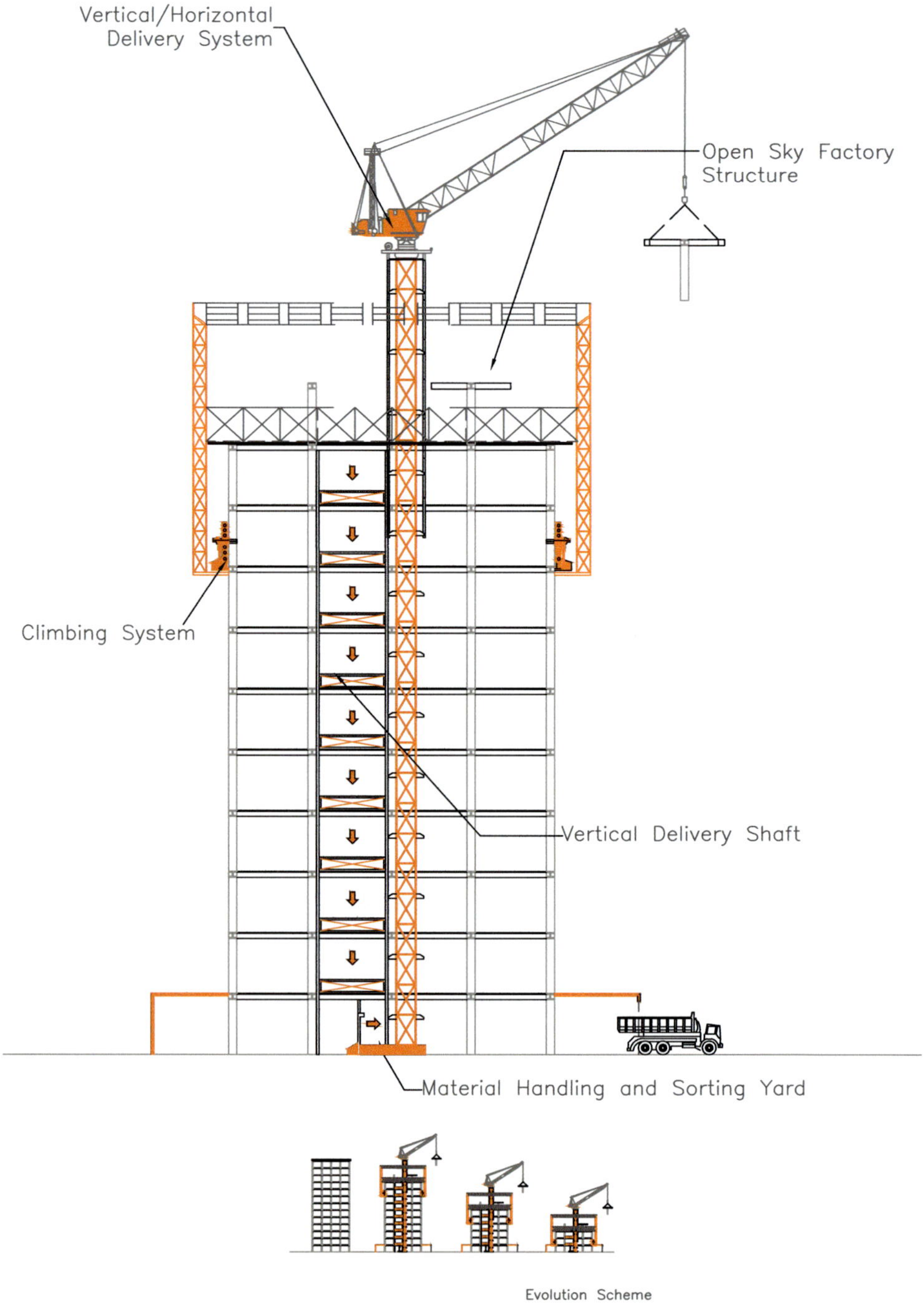

Reverse Construction Method (RCM), Company: Shimizu Corporation, Japan.

open roof structure allows the tower crane to access all locations to pick up building components and parts.

Ground Plan A centrally positioned tower crane spans the work area and accomplishes the lifting up of equipment as well as the picking up of components to be disassembled.

Figure 3.32. Sky factory interior view. (Image: RIBC, 2012)

Figure 3.33. Picking up disassembled component. (Image: RIBC, 2012)

Subsystems The RCM system systematically disassembling the building on-site consists of a set of integrated subsystems:

1. *Open sky factory structure*. The SF only provides a side cover and working platforms for facade work. Due to the intended interplay with the tower crane, the final disassembly workspace on top of the building is not covered by a roof structure (Figures 3.31–3.35). The side cover provides the basis for the working environment (protects against the wind, and partially from the sun), as well as for reducing the impact on the wider environment (by preventing noise and dust).
2. *Climbing system*. The CS is attached to the outer side of the building and, via hydraulic actuation, climbs the facade. As the system can climb up the facade, the set-up of the CS and the assembly of the factory side cover can be done conveniently on the ground, simplifying the set up and disassembly of the SF.
3. *Vertical/horizontal delivery system*. A centrally positioned tower crane can be lowered floor by floor in parallel with the SF, following the operational sequence. The working space of the tower crane allows access to disassembled components and allows those to be lowered from the SF to ground.

Figure 3.34. Sky factory interior view. (Image: RIBC, 2012)

Figure 3.35. Picking up disassembled component. (Image: RIBC, 2012)

4. *Vertical delivery shaft.* Deconstructed components can be lowered into the MHSPY through a vertical delivery shaft.
5. *Material handling, sorting, and processing yard.* The disassembly of components, material sorting, and preparation for transport are all done in an MHSPY, which is integrated into the ground floors of the building.

End-effectors The V/HDS (tower crane) is equipped with multipurpose conventional crane end-effectors. However, to guarantee a precise disassembly, new methods for on-site water and laser cutting are being applied and tested.

System Variations The installation of the SF and the CS do not require synchronization with the building's grid of columns. The subsystems can be installed to almost any building that enables the roof structure to be used as a platform for mounting the described subsystems. As the elements of the SF are orthogonal, the building should be orthogonal as well to avoid major adjustments. The facade of the building facing deconstruction should be stable enough for the application of the CS.

Robot-Oriented Design To date, no building code or guidelines exist, which, on being applied, would simplify the use of RCM.

3.2.3 Quakeproof, Quiet, Quick and Block-by-Block Building Disassembly (Cube Cut Method)

Company: Obayashi Corporation, Japan
Category: Deconstruction – OSF supported by building (moving downwards)

Evolution Scheme The Cube Cut Method SF (Figure 3.36) does not fully cover the workspace, providing only a movable, self-climbing and noise mitigating protection and scaffolding ring around the top floors, where disassembly work is done. The OSF can be erected on the ground and then lifted to the top by the climbing mechanism, where the disassembly operation starts. One or two tower cranes are erected besides the building and lift material/equipment in/out of the OSF. A newer version of the system allows addition of a light roof structure to the OSF to completely close the working environment. The roof can be opened and closed for logistics purposes but does not allow the installation of OMs.

Elevation In contrast to other systems, the Cube Cut Method does not work with overhead cranes attached to the SF, but integrates one or two tower cranes erected beside the building. The open roof structure allows the tower crane(s) to access all locations to pick up disassembled components. Disassembly work proceeds on several floors in parallel. On the uppermost floor, structural elements are cut off and delivered to the ground by the tower cranes. On the floors below this operational floor, interior finishing is disassembled and taken down by the automatic lifts.

Figure 3.36. Open sky factory moving downwards. (Image: Obayashi)

Ground Plan Strategically positioned tower cranes spans the work area and accomplish the lifting up of equipment, as well as the picking up of components to be disassembled. On the top floor, the building is cut into blocks suitable for vertical logistics (lifting). On the ground, those blocks are then further disassembled to smaller blocks or raw materials that can be recycled or reused.

Subsystems The Cube Cut system, which systematically disassembles the building on-site, consists of a set of integrated subsystems:

1. *Open sky factory structure*. The SF only provides a side cover and working platforms for facade work. Owing to the intended interplay with the tower crane, no final disassembly workspace on top of the building is covered by a roof structure. The side cover provides the basis for the working environment (protects against the wind and partially from the sun), as well as for reducing the impact on the wider environment (prevents noise and dust). A newer version of the system allows addition of a light roof structure that can be opened and closed for logistics purposes (Figures 3.38–3.41).
2. *Climbing system*. The CS is attached to the outside of the building and, via hydraulic actuation, climbs up the facade. As the system can climb up the facade, the setup of the CS and the assembly of the factory side cover can be done conveniently on the ground, simplifying the set-up and disassembly of the SF.
3. *Vertical/horizontal delivery system (tower cranes)*. Depending on the size of the building, one or two tower cranes are erected beside the building so that their workspace covers the whole floor area (Figure 3.37).

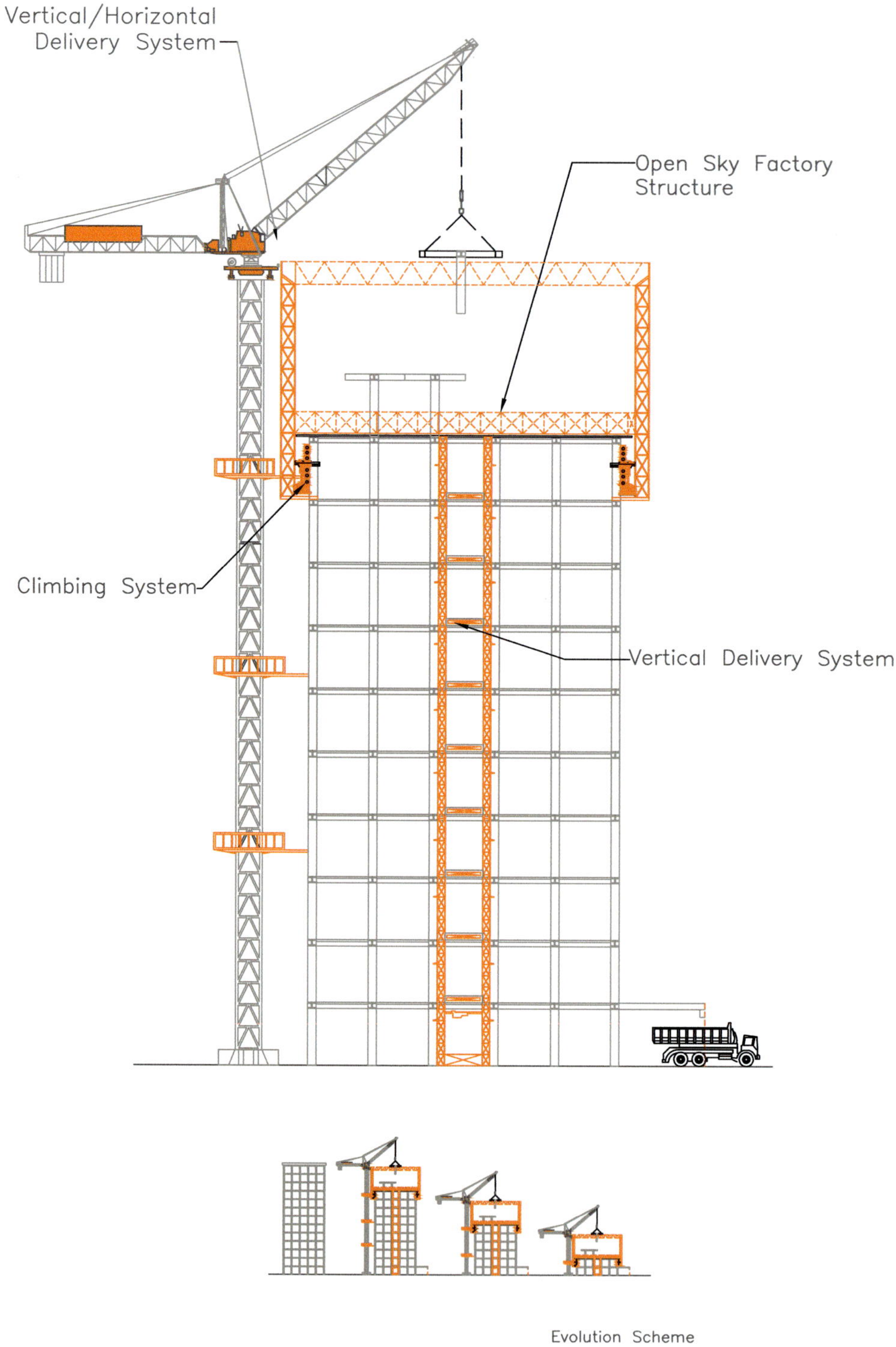

Quakeproof, Quiet Quick and Block-by-Block Deconstruction (Cube Cut Method), Company: Obayashi Corporation, Japan.

Figure 3.37. Cutting off and picking up of component in an open sky factory. (Image: RIBC, 2012)

Figure 3.38. Sky factory (temporary, light roof cover opened). (Image: Obayashi)

Figure 3.39. Sky factory top view. (Image: Obayashi)

4. *Vertical delivery system (automatic lift)*. Obayashi uses adjusted versions of the automatic lifts used within ABCS, Hybrid-ABCS, decentralized ABCS, and Big Canopy for the lowering of disassembled parts and low-level components. Larger components are lifted down by the tower cranes.
5. *Lightweight equipment*. Lightweight, conventional construction machines can be used on the operational floor.

Figure 3.40. Sky factory (temporary, light roof cover closed). (Image: Obayashi)

Figure 3.41. Sky factory top view. (Image: Obayashi)

End-effectors The V/HDS (tower cranes) is equipped with multipurpose conventional crane end-effectors. However, to guarantee a precise disassembly, new methods for on-site water and laser cutting are applied and tested.

System Variations The installation of the OSF and the CS do not require synchronization with the building's grid of columns. The subsystems can be installed to almost any building that allows the roof structure to be used as a platform for mounting the described subsystems. As the elements of the SF are orthogonal, the building should be orthogonal as well, to avoid major adjustments. Owing to the use of tower cranes for pick-up and delivery to the ground, the creation of vertical shafts within the floor is not necessary.

Robot-Oriented Design To date, no building code or guideline exists, which, on being applied, would simplify the use of the Cube Cut Method.

3.3 Ground Factory (fixed place) and Building Lowering

To date only one deconstruction system has been developed in this category. Systems in this category can be considered as a reversion and simplification of the principle used in automated construction systems that use a GF in combination with a push-up of the floors constructed on the ground. Similarly, the deconstruction systems in this category are based on the idea of installing the main factory at the bottom of the building and to concentrate the main disassembly activities on the 5–10 floors that are closest to the ground. The building structure is, therefore, disassembled at the bottom and the building is lowered by computer-controlled (hydraulic) presses, step by step. Compared to the previous categories, this system is faster (Kajima deconstructed three high-rise buildings in the middle of Tokyo within 3–4 months using the method) and the recycling rate achieved is higher (more than 90%). The higher deconstruction speed is due to the fact that the whole building serves as a factory shelter and that, in the 5–10 lower floors, disassembly can be prepared simultaneously so that when the materials reach the ground, the floors can be transported away JIT and JIS.

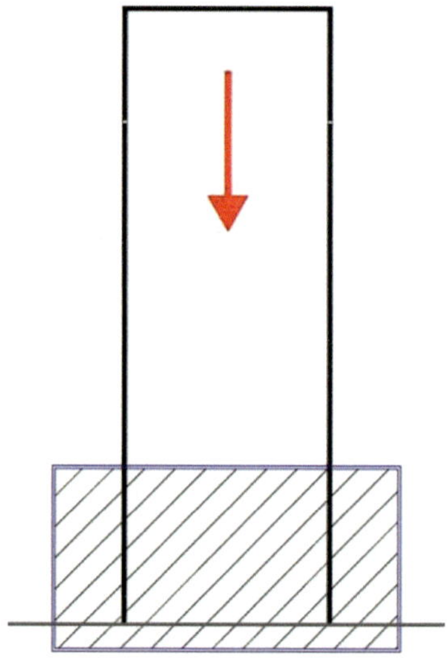

	System name	Company
30	Cut and Take Down Method "Daruma Otoshi" (DARUMA)	Kajima

However, the method also has major disadvantages. Owing to a limitation of the power of the hydraulic presses, this method cannot be applied to super high-rise buildings. Furthermore, the lowering mechanism requires the bottom or basement configuration of the building to be suitable for the installation of the (relatively heavy and bulky) hydraulic presses, and that the building is based on a certain building typology (central core combined with an outward column structure, main structure at best made of reinforced concrete). These requirements (height, bottom/basement configuration, building typology, and building material) considerably limit the application scope of the method compared to other methods (which can basically be applied to any building).

3.3.1 Cut and Take Down Method "Daruma Otoshi" (DARUMA)

Company: Kajima Corporation, Japan
Category: Deconstruction – GF (fixed place) and building lift-down

Figure 3.42. On ground disassembly of lowered and prepared floors. (Image: Kajima)

Evolution Scheme The Kajima Cut and Take Down method (DARUMA) acquired its name from the traditional Japanese game Daruma Otoshi. In this Japanese traditional game, several wooden blocks are placed on top of each and a player has to disassemble those wooden blocks, starting from the bottom by using a hammer without causing the stacked blocks to fall apart. Similarly, the DARUMA system represents a computer-controlled lowering of the building and a systematic and logistically optimized disassembly of parts and components in a GF and thus in an SE near the ground (Figure 3.42).

Elevation and Ground Plan The first time Kajima applied its DARUMA method was to disassemble the high-rise office blocks of its headquarter in Tokyo. It was selected as the initial project because it provided a systematic column grid as well as basement floors suitable for installing the lowering system. In general, the system uses 8–10 floors near the ground, including the basement, as working floors (*n*th floors in which work as part of the floor disassembly cycle takes place). Similar to a factory where a dedicated work process is performed at each workstation, on each working floor dedicated disassembly tasks are performed (disassembly of components from the building's main structure, disassembly of components into parts, sorting of material, etc.). Accordingly, the factory layout is customized for each floor on a ground plan level.

Subsystems and End-effectors The Kajima Cut and Take Down method (DARUMA) consists of a set of integrated subsystems:

1. *Core wall in combination with load transferring frame.* A reinforced concrete structure has to be placed in the centre of the buildings to be disassembled to guide the building during lowering along a frame and to serve as stabilization in case of an earthquake (Figure 3.46). The core wall can be based on an existing concrete core or a core erected only temporarily for the deconstruction period.
2. *Ground factory.* The system utilizes the 8–10 existing floors nearest the ground of the building, including the basement, as a GF. In the floors above the ground floor disassembly takes place in an SE on the still weatherproof floors. On the ground level, the other walls are disassembled so that the components disassembled in the upper floors can be removed. The ground floor is thus the main working and logistics floor (Figures 3.43–3.45, and 3.50–3.55). The lowering mechanism is installed in the basement floors.
3. *Vertical delivery system.* System for delivering tools, workers, and equipment to the 8–10 working floors and for removing disassembly waste and small parts from the working floors.
4. *Lowering system.* The lowering system consists of compact, mobile, and computer-controlled hydraulic jacks (Figure 3.47) that can be installed below the columns in the basement and equipment that allows removal of column segments (segment by segment) in the basement.
5. *Real-time monitoring and management system.* The automated operation of the lowering system as well as the overall progress of construction are monitored and controlled from an on-site control room (Figures 3.48 and 3.49).

System Variations As the system does not set up a dedicated SF or GF structure, and as the lowering system is mobile and can be placed flexibly under a column independent from the column grid structure, the system is highly flexible as far as the ground plan of the building is concerned. However, the application of the system is at present limited to a certain type of building (buildings with a columns structure and an inner core) and requires that the basement floors allow the installation and operation of the lowering system.

Robot-Oriented Design To date no building code or guideline exists, which, on being applied, would simplify the use of DARUMA. One of the key aspects of the system that can be foreseen and supported by the design of future buildings is the configuration of the basement floor, which should allow for a simplified installation and use of the lowering system. Furthermore, the column structure of the building as well as its central core could be optimized for the application of the system.

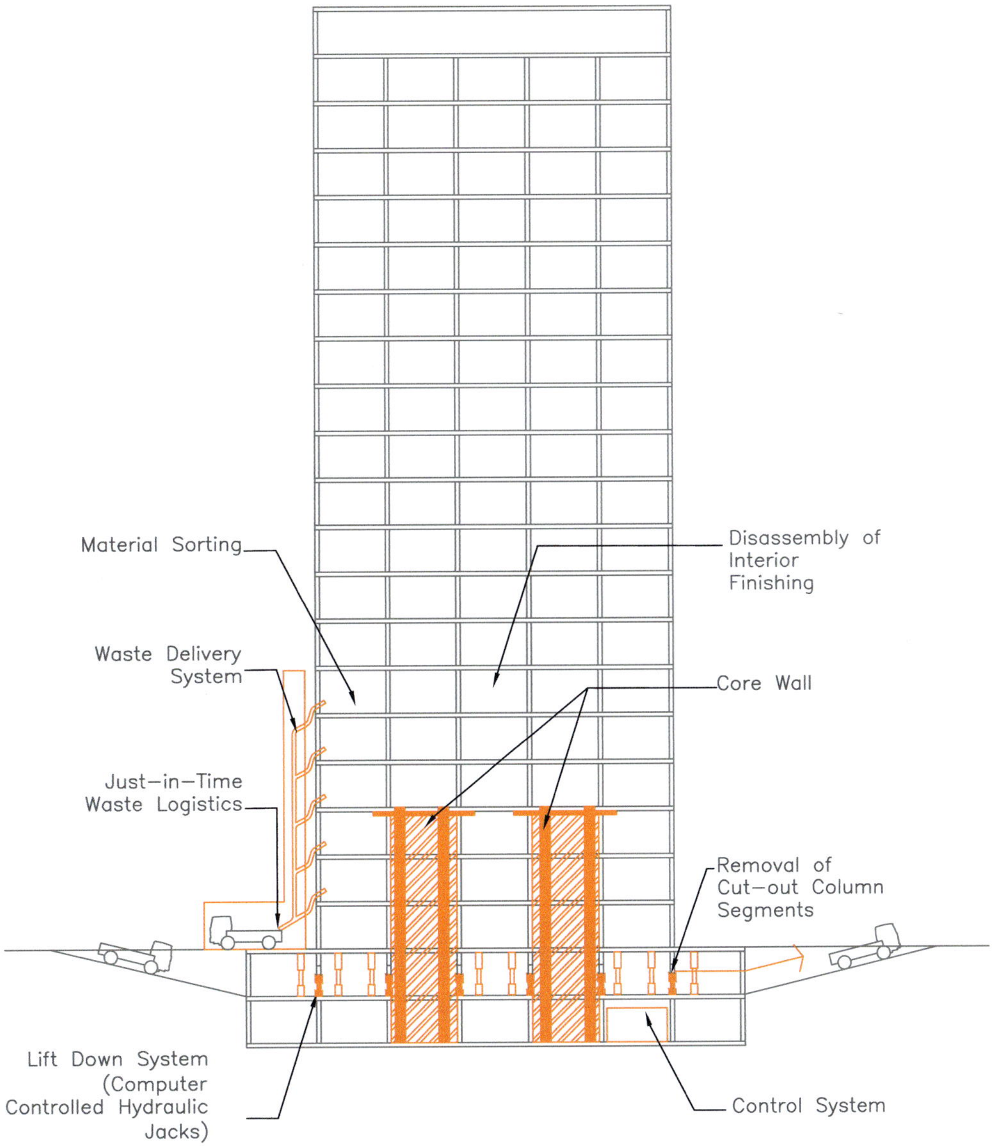

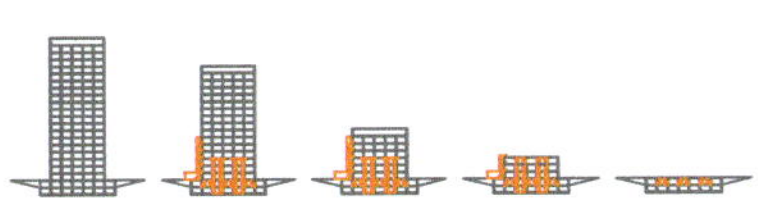

Evolution Scheme

Cut and Take Down Method "Daruma Otoshi" (DARUMA), Company: Kajima Corporation, Japan.

Deconstruction Project 2008

Figure 3.43. Disassembly by ground factory (early deconstruction stage).

Figure 3.44. Disassembly by ground factory (late deconstruction stage).

Figure 3.45. Top view on nearly deconstructed building.

Figure 3.46. Core wall and load transferring frame.

Figure 3.47. Hydraulic jacks.

Figure 3.48. Control room.

Figure 3.49. Control room.

Deconstruction Project 2012

Figure 3.50. Building during deconstruction.

Figure 3.51. An enclosing structure covers floors in which a disassembly operation takes place.

Figure 3.52. View into ground floor.

Figure 3.53. Jigs, detailed view.

Figure 3.54. Jigs guide the columns into the basement where they are cut.

Figure 3.55. Facade elements and interior finishing is removed on the ground level.

4 Conclusion: Discrepancy between Technical Capability and Efficiency

The analyses of automated/robotic on-site factories deployed so far reveal an interesting discrepancy between the technical performance achieved (potential capability of overall systems or individual subsystems) and the improvements in productivity, economic performance, and efficiency achieved. Whereas from a technical viewpoint the developed and deployed technologies (e.g., in terms of modularity, flexibility, variability, robot-oriented design [ROD]) have reached an outstanding level, efficiency, productivity, and economic performance stayed still behind the achievements in other comparable industries.

The conducted analyses claim to have identified and analysed all approaches to automated/robotic on-site factories that were conducted so far. Considering the analysed systems, and the fact that some of the systems were used several times (e.g., ABCS and SMART each up to 10 times), that subsystem applications are frequently used (e.g., Obayashi), and that currently the application of the on-site factory approach for deconstruction is taking off in Japan, it can be said that the automated/robotic on-site factory approach up to today was applied more than 60 times worldwide.

On the basis of the technical analysis (analysis of technical aspects, configuration of systems, etc.), the categorization into 13 categories shows that automated/robotic on-site factories can be installed at various locations on the construction site (on the ground, on top of buildings) and can progress in various directions (e.g., vertically upwards or horizontally), thus allowing solutions for almost any building typology (e.g., various high-rise building typologies, condominiums, point block buildings, steel buildings, concrete buildings, see Chapter 2, Sections 2.1 to 2.10). Besides construction purposes, the on-site factory approach can also be used to deconstruct buildings of different typologies (see Chapter 3, Sections 3.1 to 3.3). This means from a technical point of view, that automated/robotic on-site factories, on the basis of the applied and analysed technologies (subsystems, end-effectors, factory layouts), hold the potential to be developed for manufacturing any vertically or horizontally oriented building typology.

The analysis of subsystems and end-effector technology showed that automated/robotic on-site factories were in most cases developed as modular kits that allowed through a combination of inbuilt flexibility and modular flexibility for high variability. Each system could be broken down into multiple subsystems (see analysis of subsystems for each system in Chapters 2 and 3). Systems that work besides a main factory (e.g., sky factory [SF]) with a sub-factory (ground factory) are composed of up to 12 subsystems. Simpler systems such as, for example, the *Bauhelling* are composed of only about four subsystems. On average, systems with only one main SF or ground factory (GF) are composed of about eight subsystems. Subsystems that can be found in some form in any system are a type of factory cover, a climbing, sliding, or push-up system, a vertical and/or horizontal delivery systems (HDSs), and an ICT system for scheduling, monitoring, optimization, or equipment operation. Concerning the end-effectors, two different major strategies can be identified (see analysis of end-effectors for each system in Chapters 2 and 3). End-effectors are designed either multipurpose with inbuilt flexibility or exchangeable-modular. In the case of the modular approach up to eight end-effectors could be identified for an individual system (ABCS).

Like advanced manufacturing environments in the general manufacturing industry, automated/robotic on-site factories, generally speaking, were developed as sets of recombinable subsystems with inbuilt (robotic) flexibility or with modular flexibility (e.g., end-effector change) that can be fully synchronized with the building's modular structure through ROD. Likewise, the analysis of the variability and of applied ROD methods shows that on-site factories can be adapted within a certain extent (e.g., within a typology for which they are developed; see analysis of variability and application of ROD for each system in Chapters 2 and 3) to various building projects and thus used to automate the subsequent construction of customized, differently designed and shaped buildings. Although organizationally this capability was not yet fully used, technically speaking the analysis showed that the capability is embedded in the basic modular approach most systems built their sub-systems on.

The analysis of the evolution scheme (general approach), section ground plan (manufacturing layout), and the manufacturing process showed that the factory layouts and the optimization of the material flow from material delivery to the site to the installation in a just in time and just in sequence–like manner were a recurring theme (see analysis of factory layout and manufacturing process for each system in Chapters 2 and 3). Although some systems worked with temporary on-site warehouse modules or ground-based component processing yards (e.g., Akatuki 21, AMURAD, T-Up) a major focus was on optimizing factory layouts to erase the necessity of storage and allow a direct pick-up link between pick-up from delivery truck and assembly operations (e.g., Roof-Robo, BAM, System Skanska) and if possible even parallelized pick-up–assembly operations (for example SMART). The factory approach and the networking of subsystems to create a continuous material flow on-site in a structured environment (SE) were a direct result of the negative evaluation of the approaches to single-task construction robots (STCRs) in the 1990s (see **Volume 3**). It can be said that automated/robotic on-site factories, although they are not yet efficient enough, have definitely overcome the stand-alone technology phase and showed that the setting up of flow-line–like or production-line–like SEs on-site is possible from a technical point of view.

The analysis of productivity, efficiency, and economic performance of automated/robotic on-site factories deployed to the present day have shown that improvements were most obvious in the categories of erection speed, work productivity, and resource efficiency. Considering the fact that in conventional construction usually (of course depending on the individual use case and capabilities) a maximum of two floors can be erected in a week (not including the installation of the facade, facade sealing, installation of floor elements, and basic interior finishing, as in case of the integrated automated sites) the erection speed for standard floors increased twofold to fourfold. Work productivity and resource efficiency were improved each by more than the twofold. Furthermore, automated sites were able to reduce the amount of construction defects, safety risks, injuries, and unhealthy physical strain to near zero. The efficiency analysis showed that integrated automated construction sites – nevertheless all systems were subject to experimentation and development and did not yet fully exploit the inbuilt potential – are able to achieve in all analysed categories efficiency improvements that exceed the maximum performance possible by conventional construction methods and also those achievable by STCRs (**Volume 3**). Furthermore, automated sites would perfectly complement component manufacturing approaches and automated prefabrication (**Volume 2**). However, compared to productivity and efficiency improvements that accompanied the introduction of advanced manufacturing technology in other industries and accounting for a much more radical performance improvement (e.g., automotive industry or tunnel boring machine [TBM] tunnelling; for detailed analysis and outline; see **Volume 1**), the improvements accomplished by automated/robotic on-site factories so far can still be considered as marginal. The findings of the efficiency analysis – including cross references to the technical analysis – can be summarized as follows:

- *Erection speed.* Prior to the set-up, on-site integrated automated site systems require – in contrast to conventional sites – a planning phase (3–6 months) where the configuration of the factory and the equipment is planned and where the factory and the building kit are (e.g., according to ROD principles) adjusted to each other. The installation of the on-site site factory usually takes around 1 month. The first set-up (e.g., FACES) can take longer (2–4 months) because of certification procedures and necessary test runs. The deconstruction of the site factory usually also accounts for 1 month. The floor production rate per month (standard floor) is between a minimum of two floors (e.g., AMURAD) and more than six floors per month (ABCS, SMART with learning effect). The floor production includes the installation of the facade, facade sealing, installation of floor elements, and basic interior finishing. The floor production rate is, apart from the speed of the systems, dependent on the size of the standard floors. Most companies worked with an initial floor erection cycle (FEC) of 6 days. Some companies (e.g., Obayashi, Shimizu) also experimented with shorter cycles (3 days and less) after the execution of several projects. An FEC usually covered parallel work on multiple floors or levels (up to four floors/levels). In general, it can be said that systems that erected steel based buildings showed a higher erection speed than systems erecting concrete-based buildings. On high-rise construction sites that use conventional construction methods, the construction of two floors per week (not including the installation of the facade, facade sealing, installation of floor elements, and basic interior finishing as in the case of the integrated automated sites) can be considered as fast.
- *Configuration and adaptability*. Companies operated equipment (OMs, HDS, VDS) used for the installation of elements (columns, beams, floor slabs, facade

elements) at speeds ranging from 10 to more than 40 m/min. Companies such as Shimizu and Obayashi that worked with a multitude of smaller OMs or travelling hoists were able to operate their equipment faster (40 m/min) than companies that focussed on larger and more powerful cranes (with higher accuracy/repeatability), such as Maeda and Goyo (20 m/min). The lowest operational speed used the positioning robots of Kajima (10 m/min), which had to process heavy precast concrete components. Besides the horizontally working positioning equipment, the vertical lifts supplying those systems play a crucial role in creating a continuous material flow. In larger projects, companies often used several lifts in parallel (e.g., Obayashi). Companies experimented with different automation ratios. The positioning of components such as columns could be done automatically, half automatic (e.g., automation of positioning operation to near the installation location followed by manual final positioning), and manually via remote control. On the one hand, owing to the processing of sensor and positioning data, full automation turned out to be 5–20% slower than manual operation. On the other hand, full automation saved person-hours and personnel.

Companies also adapted the configuration of the systems from project to project to the individual specifications of the building. Therefore the allocation of, for example, OM systems, VDS, HDS, trolley hoists, and lifts as well as their number was varied, forming variations of the factory layouts of the SFs and GFs. In general it can be said that companies such as Shimizu that used a fine-grained OM system with smaller modules had more possibilities to adjust and reconfigure their systems to be used in a new project than companies that used large and powerful systems, such as, for example, Goyo. Experiments with the number of workers on-site showed that on automated sites the optimum input/output ratio according to the optimum principle is achieved with very low numbers of workers working at the same time on-site.

- *Productivity.* The reduction of human labor required on automated construction sites compared to conventional construction sites (comparable projects executed by the reference companies) was between 20% and 70%. The highest reduction of labour was achieved in the fields of facade installation (steel-based buildings), concrete and steel bar/formwork work (concrete-based buildings), and safety-related temporary installations. In addition, the workforce structure was considerably altered and shifted from simple crafts-oriented work (reduction of workers in that field) more to work that needs multiskilled workers capable of performing various tasks on-site, and supervisory workers capable of supervising and controlling the new equipment. Automated construction sites require about 11–15 workers constantly on-site, which equals the number of workers on-site when operating a TBM. Furthermore, learning effects on automated construction sites (e.g., time necessary for installation of components with equipment, welding, etc.) were enormous and accounted for up to 50% within one project. So, for example, the installation of a reinforced concrete column by AMURAD was reduced from about 20 to 10 minutes and the installation of a steel column by SMART from 40 to 20 minutes. However, during the deployment of their first sites, Shimizu and Obayashi had the intention of training the workforce in general in the use of the new technology (according to Japanese philosophy, knowledge about new technologies and tools has to be spread quickly among the work force by special training procedures) and thus exchanged half of the workforce from project to project with new workers, to train as many workers as fast as possible in using the new technology. This mitigated the learning effect cross projects considerably and it can be assumed that the aforementioned floor erection speed could be much higher when exploiting learning effects across projects.

- *Resource efficiency*. The positive impact of integrated, automated construction sites was mentioned by almost all companies. However, only one company (Shimizu) made a detailed and valid evaluation. It found out that automated sites reduce the amount of construction waste generated on-site by 70% owing to the high degree of prefabrication and the on-site material management system. This is equal to a reduction of the waste volume by about 700 tons. Of course, waste was also generated in prefabrication factories; however, controlled environments and material flows in off-site environments generally make it much easier to recycle or reuse these wastes. The companies Sekisui Heim and Toyota Home, for example, which prefabricate houses in large scale in Japan, operate certified zero-waste factories.
- *Quality, health, and safety.* All companies used material management and site control systems that allowed them to monitor construction activity, construction quality, and the progress of the construction in real time. Also most companies developed simulators that allowed simulating and optimizing the set-up or configuration of the system on-site and its operation. Shimizu even developed a system for simulating and optimizing the disassembly of the factory and the machine systems on-site. Furthermore, companies analysed – by simulations and by measurements – the impact of the factory environment on the working conditions (safety physical strain, light conditions, etc.). The working conditions enhance not only safety and health but also productivity and (as workers can then better concentrate on their work task) the construction defect rate. The environment reduced the physical strain (e.g., measured by heart rate) considerably as well as the risk of falls during facade work or the dropping of material/injury during transport.
- *Usability*. Usability studies showed that generally the new factory-like environment was neither evaluated positively nor negatively by workers themselves. An overall positive evaluation by workers was mainly mitigated by the fact that the systemized work environment obviously reduced the room for unnecessary breaks and laid back working attitude, changed the habits of the construction workers, and thus was not in all aspects evaluated positively. This phenomenon accompanies many new developments that involve technologies that are able to systemize, track, or record people and processes. Aspects that are related to the improvement of the work environment, working conditions, and work flow were evaluated positively by the workers.

Automated/robotic on-site factories state the probably most complex approach towards automated construction developed so far. Both the development of the specialized, automated equipment (overhead manipulators, climbing systems, factory structures, end-effectors, etc.) used, as well as the integration of this equipment to interlinked, temporary factories that allow coherent, uninterrupted material flow in SEs on the construction site are challenges that other approaches to construction automation do not face. However, it can also be said that once these challenges are sufficiently addressed, automated/robotic on-site factories would in terms of both technical and efficiency/economic performance significantly outperform the other approaches. Automated/robotic on-site factories are a very complex form of innovation that, represented by the S-curve model (see also **Volume 1, Section 1.2**), probably will prevail longer in the research and innovation phase than other approaches, but in the end also will allow for much higher limits in terms of technical and economic performance.

References

ABCS (1993) Riverside Sumida Bachelor Dormitory Documentary Video [VHS]. Tokyo: Obayashi Corporation.

Bock, T. (1988) *A Study on Robot Oriented Construction and Building System*. Dr.-Ing. dissertation, Faculty of Engineering, Chair of Building Production, The University of Tokyo.

Bock, T. (2009) "Urban Mining – der nachhaltige Baubetrieb: Vom automatisierten Auf- zum Rückbau". In *Festschrift TU Graz zum 40jährigen Jubiläum des Institutes für Baubetrieb und Bauwirtschaft*. Graz, Austria: Institutes für Baubetrieb und Bauwirtschaft.

Bock, T., Linner, T., Miura, S. (2011) "Robotic high-rise construction of pagoda concept: Innovative earthquake-proof design for the Tokyo Sky Tree". *Proceedings of Council for Tall Buildings and Urban Habitat (CTBUH) World Conference*, Seoul, Korea: CTBUH.

Chu, B., Jung, K., Ko, K. H., Hong, D. (2010) "Mechanism and Analysis of a Robotic Bolting Device for Steel Beam Assembly". In *Proceedings of International Conference on Control, Automation and Systems 2010*, pp. 2351–2356. Gyeonggi-do, Korea.

Chu, B., Jung, K., Chu, Y., Hong, D., Lim, M., Park, S., Lee, S., Kim, M. C., Ko, K. H. (2009) "Robotic automation system for steel beam assembly in building construction". In Proceedings of 4th International Conference on Autonomous Robots and Agents, ICARA, pp 38–43. Wellington, New Zealand.

Cousineau, L., Miura, N. (1998) *Construction Robots: The Search for New Building Technology in Japan*. Reston: American Society of Civil Engineers (ASCE).

Fujii, T., Hagiwara, C., Morita, M., Miyaguti, M. (1995) "Development of roof push up construction method". In Budny, E., McCrea, A., Szymanski, K. (eds.), *Automation and Robotics in Construction*, Vol. 7, pp. 193–201. Amsterdam: Elsevier.

Gotthelf, F. (1960) *Ernst Neufert. Ein Architekt unserer Zeit*. Frankfurt/Berlin: Verlag Ullstein.

Goyo/Penta Ocean (2001) FACES Documentary Video [VHS]. Tokyo: Goyo/Penta Ocean.

Hasegawa, Y. (1985) "Robotization of reinforced concrete building construction in Japan". In *Proceedings of the 3rd International Symposium on Automation and Robotics in Construction*, pp. 423–433. Marseille, France: International Association for Automation and Robotics in Construction.

Hasegawa, Y. (1999) *Robotization in Construction*. Tokyo: Kogyo Chosakai Publishing.

Hasegawa, Y., Onoda, T., Tamaki, K. (1992) "Modular robotic application for flexible building construction". In *Proceedings of the 13th International Symposium on Automation and Robotics in Construction*, pp. 113–122. Tokyo, Japan International Association for Automation and Robotics in Construction.

Ikeda, Y., Harada, T. (2006) "Application of the automated building construction system using the conventional construction method together". In *Proceedings of the 23rd International Symposium on Automation and Robotics in Construction*, pp. 722–727. Tokyo, Japan: International Association for Automation and Robotics in Construction.

Itou, M., Nishita, K., Houkyou, M. (1994) "Application of all-weather automatic building construction systems". Japan: Construction Robot Symposium.

Kajima (2000) AMURAD Documentary Video [VHS]. Tokyo: Kajima.

Kajima (2009) Documentary Picture Set [JPEG]. Tokyo: Kajima Corporation Research Department.

Kajima (2011) Interview with Kajima R&D staff at ISARC 2011. Seoul, Korea, June.

Kang, T., Nam, C., Lee, U., Doh, N., Park, G. (2011) "Development Of Robotic-Crane Based Automatic Construction System For Steel Structures Of High-Rise Buildings". In *Proceedings of the 28th International Symposium on Automation and Robotics in Construction*, pp 670–671. Seoul, Korea: International Association for Automation and Robotics in Construction.

Korea University (2010) Documentary Video Robotics and Crane based Automated Construction System (RCACS) for High-rise Buildings [Windows Media Video]. Seoul: Korea University.

Kress, C. (2008) *Zwischen Bauhaus und Bürgerhaus. Die Projekte des Berliner Bauunternehmers Adolf Sommerfeld*. PhD dissertation. Technische Universität Berlin.

Kress, C. (2011) *Adolf Sommerfeld – Andrew Sommerfield*. Berlin: Lukas Verlag.

Lee, S., Pan, W., Linner, T., Bock, T. (2015) "A framework for robot assisted deconstruction: Process, sub-systems and modelling". In *Proceedings of the 32nd International Symposium on Automation and Robotics in Construction and Mining (ISARC2015)*, Oulu, Finland, 15–18 June, 2015.

Linner, T. (2013) "*Automated and Robotic Construction: Integrated Automated Construction Sites*." Dr.-Ing. dissertation, Technische Universität München – categorization and analyses in this volume are based on chapter 5 of the thesis.

Maeda, J. (1994) "Development and application of the SMART system". In *Automation and Robotics in Construction*, Vol. 6, pp. 457–464. Oxford: Elsevier.

Maeda, J., Miyatake, Y. (1997) "Improvement of a computer integrated and automated construction system for high-rise building and its application for RC (Rail City) Yokohama building". In *Proceedings of the 14th International Symposium on Automation and Robotics in Construction*, pp. 139–146. Pittsburgh, PA: International Association for Automation and Robotics in Construction.

Miyakawa, H., Ochiai, J., Oohata, K., Shiokawa, T. (2000) "Application of automated building construction system for high rise office building". In *Proceedings of the 17th International Symposium on Automation and Robotics in Construction*, pp. 1–6. Taipei, Taiwan: International Association for Automation and Robotics in Construction.

Morita, M., Muro, E., Kanaiwa, T., Nishimura, H. (1993) "Study on simulation of roof pushup construction method". In G. H. Watson, G. H. Tucker, R. L. Walters (eds.), *Automation and Robotics in Construction*, Vol. 5, pp. 1–8. Amsterdam: Elsevier.

Moselhi, O., Khan, Z. (2012) "Significance ranking of parameters impacting construction labor productivity". *Construction Innovation: Information, Process, Management*, 12(3): 272–296.

NCC (2006) *Annual Report 2006* [Research Report]. Sweden: NCC Sweden.

NCC (2012) Documentary picture set NCC Komplett [JPEG]. Solona: NCC Communication/ Press Department.

Neufert, E. (1943) *Bauordnungslehre*. Berlin: Volk und Reich Verlag.

Nishimatsu (2012) "Move hat implementation" [Research Report]. Japan: Nihimatsu Corporation.

Obayashi (1993) Riverside Sumida Bachelor Dormitory Documentary Video [VHS]. Tokyo: Obayashi Corporation.

RIBC (2012) "Study of building cost". *The Seasonal Bulletin*, No. 79. Tokyo: Research Institute on Building Cost (RIBC).

Sakamoto, S., Kumano, T. (1991) "Research and development of totally mechanized construction system for high-rise buildings". In *Proceedings of the 8th International Symposium on Automation and Robotics in Construction*, pp. 197–206. Stuttgart, Germany: International Association for Automation and Robotics in Construction.

Sakamoto, S., Mitsuoka, H. (1994) "Totally mechanized construction system for high-rise buildings (T-UP system) ". In *Proceedings of the 11th International Symposium on Automation and Robotics in Construction*, pp. 465–472. Brighton, UK: International Association for Automation and Robotics in Construction.

Sakamoto, S., Yokuda, T. (1994) "Application for totally mechanised construction system for high-rise buildings". In *Proceedings of 8th Construction Robot Symposium*. Tokyo: Architectural Institute of Japan (AIJ).

Sekiguchi, T., Honama, K., Mizutani, R., Takagi, H. (1997) "The development application of an automatic building construction system using push-up machines". In *Proceedings of the 14th International Symposium on Automation and Robotics in Construction*, pp. 321–328. Pittsburgh, PA: International Association for Automation and Robotics in Construction.

Shibata, T. (1999) "Study on robotics application for construction works". In Hasegawa, T. (ed.), *Robotization in Construction*, Tokyo.

Shimizu, "The tower smart application". www.shimz.co.jp/news_release/2000/357.html (Accessed 24 November 2012).

Shimizu (2013) Documentary picture set tower SMART [JPEG]. Tokyo: Shimizu Corporation Public Relations Department.

Son, H., Kim, C. (2010) "3D structural component recognition and modeling method using color and 3D data for construction progress monitoring". *Automation in Construction*, 19(7): 844–854.

Tani, Y., Nakano, M., Okoshi, M., Maeda, S., Isa, H. (1996) "Research and development of automatic system for open caisson method". In *Proceedings of the 13th International Symposium on Automation and Robotics in Construction*, pp. 323–332. Tokyo, Japan: International Association for Automation and Robotics in Construction.

Ueno, T. (1998) "Automation and robotics in construction in Japan – state of the art". In *Proceedings of the 15th International Symposium on Automation and Robotics in Construction*, pp. 33–36. Munich, Germany: International Association for Automation and Robotics in Construction.

Wakisaka, T., Furuya, N., Hishikawa, K., Inoue, Y., Shiokawa, T., Yoshino, K., Hamada, K., Inoue, F., Kurita, K. (1997) "Automated construction system for high-rise reinforced concrete buildings". In *Proceedings of the 14th International Symposium on Automation and Robotics in Construction*, pp. 111–118. Pittsburgh, PA: International Association for Automation and Robotics in Construction.

Yoshida, N., Kanagawa, T., Tani, Y., Oda, Y. (1997) "Development of an automatic-oriented Sheltered Building Construction System". In *Proceedings of 14th International Symposium on Automation and Robotics in Construction*, pp. 129–138. Pittsburgh, PA: International Association for Automation and Robotics in Construction.

Index

Printed in the United Kingdom by TJ Clays Ltd.